BRITISH RAIL WAGON FLEET

BRITISH RAILWAYS
"B" – PREFIXED SERIES FREIGHT STOCK
B125611 – B955247

piled by
ickenson

GW00467700

PUBLISHING

Published by:-
South Coast Transport Publishing
Hampshire, England

Printed by Itchen Printers Ltd.
Southampton. Tel: (0703) 227161

ISBN 1 872768 11 3

Although this particular example was withdrawn prior to 1990, it was felt inappropriate for this publication not to include a picture of a 21T Mineral Wagon. MDV, B310881 is pictured at Burrows Sidings in March 1988.   Peter Ifold

In March 1988 Burrows Sidings also held DB340140, ZDV, a 25t General Materials Wagon prior to its transfer to departmental service.
                                                              Peter Ifold

British Rail Wagon Fleet

B.R. ("B" - Prefixed Series) Freight Stock (B125611 - B955247)

Introduction

In 1948 on the formation of British Railways a large number of wagons within different number series which often overlapped were inherited from the pre-nationalisation companies and private owner fleets acquired pre-1948. To differentiate the various existing number series, existing stock was prefixed by a letter denoting its source company e.g. "S" indicated stock from the Southern Railway (SR). An additional number series was introduced by British Railways for new stock and this carried the prefix "B". This number series was continued by BR until the 1960's when air braked stock was developed in preference to vacuum braked stock and a new number series was introduced without a prefix.

This publication which is a companion volume to the other publications on British Rail Wagon Fleet records all wagons in the "B"-prefixed number series remaining extant on BR records at 1st January 1990. A significant number of wagons have been withdrawn since that date and in accordance with standard practice withdrawn vehicles are indicated by (W). The number range in the title reflects the stock at 1st January 1990 rather than the full range of the series which was wider prior to the many disposals. Wagons in the series 350000 - 355396, 511000 - 511022, 601001 - 601493, 601999, 602001 - 602951 and 873978 - 874185 which originally entered service with a "B" prefix have not been included in this publication as these now form part of the air braked number series and have been included in our publication on air braked freight stock. It was also decided to exclude the two Bullion Vans, 889301 and 889302, as these were not "B"-prefixed.

Wagons which have subsequently been transfered to the Engineering Sections of BR have prefix letters applied to identify the "owning" section. All engineering service stock carry a D prefix in addition to a regional letter which identifies for whom the stock was built e.g. DB = British Railways stock. A third letter is added at the front of the prefix to identify the "owning" section. The "owning" prefixes appropriate to this publication are as follows:-

```
A = Director of Mechanical & Electrical Engineering
C = BREL
K = Director of Signal & Telegraph Engineering
L = Director of Mechanical & Electrical Engineering)
        (BRB HQ Electrification Projects)
R = Director of Research
T = Director of Operations
The addition of no third prefix letter = Director of Civil
Engineering
```

TOPS (Total Operating Processing System) was introduced by BR in 1972 and each vehicle was allocated a three letter TOPS code. In respect of codes allocated to stock included in this publication the following letters apply:-

```
1. First letter = main type group
    B  Bogie Steel
    C  Covered Bulk
```

```
CA Brake Vans (Traffic Department)
F   Flat Wagons
H   Hoppers
M   Mineral Wagons
R   Railway Operating
S   2-Axle Steel
Y Service, bogie freight
Z   Service, 2-axle freight
```

2. Second letter = Sub-division of main type
3. Third letter = Brake type

```
A   Air Brake
B   Air Brake, through vacuum pipe
O   No Brake (Hand Only), unfitted
P   No Brake (Hand Only), vacuum pipe only
Q   No Brake (Hand Only), air pipe only
R   No Brake (Hand Only), dual air and vacuum pipe
V   Vacuum Brake
W   Vacuum Brake, through air pipe
X   Dual Brake (air and vacuum)
```

Throughout this publication the TOPS and Design codes for each number series together with the Tare Weight and Gross Laden Weight (G.L.W.) are applicable to all wagons within that series unless indicated otherwise.

Although TOPS Codes have been applied generally to all wagons since 1972, the Engineering Departments have continued to use "Marine" codenames and this includes some wagons in this series which have been transfered to the Engineering Sections. Where appropriate these codenames are in the descriptions to the appropriate number series.

The description within each number series is that applicable at the time of build. Since first entry into service many wagons have subsequently been modified or rebuilt and the description may no longer be applicable. Readers should note the weights expressed in the description are imperial tons though the tare and gross laden weights reflect metric tonnes.

The compiler wishes to acknowledge the photographic contributions made by the Paul Bartlett and Peter Ifold.

John Dickenson

February 1994

Number Series: B125611 - B126860

Description: 16T Mineral End Door Wagon
Builder: Derbyshire C & W Co Ltd
Diagram No.: 1/108
Tare Weight: 8.0t
Design Code: ZH017C

Lot No.: 2385
Built: 1953
G.L.W.: 24.5t
Tops Code: ZHV

DB126447 (W)   DB126498 (W)

Number Series: B159792 - B160571

Description: 16T Mineral End Door Wagon
Builder: P & W McLellan Co Ltd
Diagram No.: 1/108
Tare Weight: 7.8t + 8.0t
Design Code: MC001D + ZH017E

Lot No.: 3143
Built: 1958
G.L.W.: 19.0t 25.5t +
Tops Codes: MCV +   ZDV

DB160073 (W)   DB160214 (W)   DB160343 (W)   DB160444 (W)   DB160539 (W)
DB160152 (W)   DB160309 (W)   DB160362+(W)   DB160537 (W)   DB160568 (W)
DB160208 (W)   DB160336 (W)

Number Series: B168352 - B170851

Description: 16T Mineral End Door Wagon
Builder: Pressed Steel Co Ltd
Diagram No.: 1/008
Tare Weight: 7.8t
Design Code: MC001C

Lot No.: 2634
Built: 1957
G.L.W.: 25.5t
Tops Code: MCV

B169190 (W)

Number Series: B261309 - B261508

Description: 16T Mineral End Door Wagon
Builder: Butterley C & W Co Ltd
Diagram No.: 1/108
Tare Weight: 8.0t
Design Code: ZH017C   ZH017E *

Lot No.: 3178
Built: 1957
G.L.W.: 24.5t
Tops Code: ZHV

DB261391 (W)   DB261429 (W)   DB261470*(W)

Number Series: B261509 - B262608

Description: 16T Mineral End Door Wagon
Builder: Cambrian Wagon & Eng Co Ltd
Diagram No.: 1/108
Tare Weight: 7.8t + 8.0t
Design Code: MC001D + ZH017E

Lot No.: 2806
Built: 1957
G.L.W.: 24.5t 25.5t +
Tops Code: MCV +   ZHV

DB261647 (W)   DB261649 (W)   DB261661 (W)   B261705+(W)

Number Series: B266209 - B267008

Description: 16T Mineral End Door Wagon
Builder: Gloucester R C & W Co Ltd   Lot No.: 2810
Diagram No.: 1/108   Built: 1957
Tare Weight: 8.0t   G.L.W.: 24.5t
Design Code: ZH017E   Tops Code: ZHV

DB266265 (W)	DB266389 (W)	DB266527 (W)	DB266691 (W)	DB266777 (W)
DB266342	DB266409 (W)	DB266669 (W)	DB266741	DB266787 (W)
DB266359 (W)	DB266471 (W)	DB266689 (W)	DB266763 (W)	DB266808 (W)
DB266365 (W)	DB266521 (W)			

Number Series: B282150 - B282769

Description: 24.5T Mineral End Door Wagon
Builder: BR (Ashford Works)   Lot No.: 3244
Diagram No.: 1/118   Built: 1959
Tare Weight: 10.5t   G.L.W.: 32.0t
Design Code: MD015M   Tops Code: MDO

 B282270 (W)   B282687 (W)   B282707 (W)

Number Series: B282770 - B282994

Description: 24.5T Mineral End Door Wagon
Builder: BR (Ashford Works)   Lot No.: 3302
Diagram No.: 1/118   Built: 1960
Tare Weight: 10.5t   G.L.W.: 32.0t
Design Code: MD015M   Tops Code: MDV

 B282847 (W)

Number Series: B290000 - B290485

Description: 25t Mineral End Door Wagon   Lot No.: 3920 and 3921
Builder: BR (Shildon Works)   Built: 1977-78
Diagram No.: 1/148   G.L.W.: 31.5t 32.0t *
Tare Weight: 9.95t # 10.2t 10.25t + 10.5t *
Design Code: MD015R MD015U + MD015V # MD015Y *
Tops Code: MDO

B290050 (W)	B290074*(W)	B290093 (W)	B290190 (W)	B290326 (W)
B290073+(W)	B290075*(W)	B290183 (W)	B290284#(W)	B290419+(W)

Number Series: B310000 - B310999

Description: 21T Mineral Wagon   Lot No.: 3387
Builder: BR (Shildon Works)   Built: 1961-62
Diagram No.: 1/119   G.L.W.: 32.7t 33.0t *+
Tare Weight: 11.2t 11.3t *   Tops Code: MDV   MDW *
Design Code: MD008C MD008D * ZD160A +   ZDV +

```
B310029 (W)    B310179 (W)    B310402 (W)    B310556 (W)    B310691 (W)
B310033 (W)    B310183 (W)    B310405*(W)    B310582 (W)    B310715 (W)
B310048 (W)    B310213 (W)    B310408 (W)    B310601 (W)    B310726 (W)
B310063 (W)    B310218 (W)    B310426 (W)    B310609 (W)    B310729 (W)
B310086 (W)    B310227 (W)    B310429*(W)    B310610 (W)    B310748 (W)
B310090 (W)    B310236 (W)    B310466 (W)    B310612 (W)    B310773*(W)
B310114 (W)    B310313 (W)    B310476 (W)    B310632 (W)    B310808 (W)
B310115 (W)    B310314 (W)    B310484 (W)    B310641 (W)    B310810 (W)
B310119 (W)    B310315 (W)    B310496*(W)    B310648 (W)    B310818 (W)
B310154 (W)    B310355 (W)    B310501 (W)    B310660 (W)    B310821 (W)
LDB310160+     B310362 (W)    B310524 (W)    B310668 (W)    B310883 (W)
B310161*(W)    B310372 (W)    B310545*(W)    B310684 (W)    B310893 (W)
B310176 (W)    B310400 (W)    B310546*(W)    B310686 (W)
```

Number Series: B311000 - B311949

Description: 21T Mineral Wagon
Builder: BR (Derby Works)
Diagram No.: 1/120
Tare Weight: 11.4t
Design Code: MD008B   MD008E *   ZD161A +

Lot No.: 3390
Built: 1961-62
G.L.W.: 33.00t
Tops Code: MDV   MDW *
ZDV +

```
B311011 (W)    B311167*(W)    B311221 (W)    B311368 (W)    B311777 (W)
B311016 (W)    B311169*(W)    B311254 (W)    B311371*(W)    B311793 (W)
B311031 (W)    B311178 (W)    B311263*(W)    B311403 (W)    B311796 (W)
B311035 (W)    B311203*(W)    B311266 (W)    B311491 (W)    B311800 (W)
B311045 (W)    B311206 (W)    B311267 (W)    B311623 (W)    B311813 (W)
B311052 (W)    B311207 (W)    B311311 (W)    B311704 (W)    B311816 (W)
B311062 (W)    B311216 (W)    B311312 (W)    LDB311717+     B311827 (W)
B311101*(W)    B311217 (W)    B311315*(W)    B311746 (W)    B311835 (W)
B311143 (W)    B311219 (W)    B311341 (W)    B311760 (W)    B311935 (W)
B311160 (W)
```

Number Series: B312000 - B313499

Description: 21T Mineral Wagon
Builder: BR (Derby Works)
Diagram No.: 1/120
Tare Weight: 11.4t
Design Code: MD008B   MD008E *

Lot No.: 3430
Built: 1962-63
G.L.W.: 33.0t
Tops Code: MDV   MDW *

```
B312020 (W)    B312569 (W)    B312916 (W)    B313139 (W)    B313349 (W)
B312046 (W)    B312588 (W)    B312923 (W)    B313180 (W)    B313382 (W)
B312121 (W)    B312618 (W)    B312950 (W)    B313194 (W)    B313384 (W)
B312197 (W)    B312703 (W)    B312987 (W)    B313195 (W)    B313387 (W)
B312211 (W)    B312731*(W)    B312990 (W)    B313219 (W)    B313393 (W)
B312249*(W)    B312734 (W)    B313000 (W)    B313260*(W)    B313425*(W)
B312275 (W)    B312768 (W)    B313014 (W)    B313273 (W)    B313452*(W)
B312317 (W)    B312778 (W)    B313035 (W)    B313282 (W)    B313466 (W)
B312323 (W)    B312794*(W)    B313042 (W)    B313309 (W)    B313474 (W)
B312456*(W)    B312864 (W)    B313056 (W)    B313311 (W)    B313480 (W)
B312465 (W)    B312896 (W)    B313065 (W)    B313337 (W)    B313492 (W)
B312474 (W)    B312902 (W)    B313075 (W)    B313348 (W)    B313499 (W)
B312520 (W)    B312907 (W)    B313111*(W)
```

Number Series: B313500 - B314499

Description: 21T Mineral Wagon
Builder: BR (Shildon Works)
Diagram No.: 1/119
Tare Weight: 11.4t
Design Code: MD008B MD008E *

Lot No.: 3438
Built: 1962-63
G.L.W.: 33.0t
Tops Code: MDV MDW *

B313528 (W)	B313618*(W)	B313811 (W)	B313986 (W)	B314186 (W)
B313557 (W)	B313628 (W)	B313849 (W)	B314042 (W)	B314240 (W)
B313575 (W)	B313675 (W)	B313874 (W)	B314080 (W)	B314324*(W)
B313580 (W)	B313682 (W)	B313912 (W)	B314084 (W)	B314325 (W)
B313593 (W)	B313778 (W)	B313975 (W)	B314086 (W)	B314421 (W)
B313598 (W)	B313793 (W)	B313983*(W)	B314124 (W)	B314489 (W)
B313606 (W)				

Number Series: B314500 - B314999

Description: 21T Mineral Wagon
Builder: BR (Derby Works)
Diagram No.: 1/119
Tare Weight: 11.2t * 11.4t 11.5t +
Design Code: MD008B MD008C * ZY127A +

Lot No.: 3439
Built: 1963-64
G.L.W.: 32.7t * 33.0t
Tops Code: MDV ZYV +

B314535 (W)	B314660 (W)	LDB314756+	B314883 (W)	B314926 (W)
B314536 (W)	B314692 (W)	B314817 (W)	B314884 (W)	B314928 (W)
B314618 (W)	B314721 (W)	B314829*(W)	B314905 (W)	B314936 (W)
B314623*(W)	B314726 (W)	B314831 (W)	B314909 (W)	B314950 (W)
B314627 (W)	B314732 (W)	LDB314842+	B314914 (W)	B314967 (W)
B314630 (W)	B314748 (W)	B314852 (W)	B314921 (W)	B314995 (W)

Number Series: B333002 - B333499

Description: 24.5T 2 Axle Coal Hopper Wagon
Builder: BR (Shildon Works)
Diagram No.: 1/148
Tare Weight: 11.0t
Design Code: HU001A

Lot No.: 2609
Built: 1955-56
G.L.W.: 35.0t
Tops Code: HUO

B333092 (W)   B333281 (W)

Number Series: B340000 - B340925

Description: 25t 2 Axle General Materials Wagon
Builder: BR (Shildon Works)
Diagram No.: 1/146
Tare Weight: 9.5t ! 10.5t
Design Code: ZD123D   ZD123E b ZD123F :
             ZD123J % ZD152A + ZD153C !
             ZD153D c HT031M * HT031N =
             HT031P - HT032M # HT032N "

Lot No.: 3916
Built: 1977-78
G.L.W.: 31.0t ^ 32.0t
        34.5t !
        35.5t :%#=a
Tops Code: ZDV b:%+!c
           HTV *=-#"

B340009*(W)	B340023*(W)	B340043*(W)	B340051*(W)	DB340066 (W)
B340018*(W)	B340039-(W)	DB340050%(W)	B340054*(W)	B340067-(W)

B340068 (W)	DB340258*(W)	B340406#(W)	B340563#(W)	B340718-(W)
B340069"(W)	B340261-(W)	B340411#(W)	B340570-(W)	DB340719#(W)
B340070#(W)	DB340266:(W)	B340415*(W)	B340571=(W)	B340727-(W)
DB340086 (W)	B340270*(W)	B340417#(W)	DB340575-(W)	DB340728 (W)
B340094#(W)	B340272#(W)	DB340418	B340580-(W)	B340737=(W)
B340098*(W)	B340273*(W)	B340423#(W)	B340581a(W)	B340739=(W)
DB340100 (W)	B340286-(W)	B340427#(W)	B340592#(W)	DB340748c(W)
B340105=(W)	B340287:(W)	B340431a(W)	B340602"(W)	B340750-(W)
DB340108*(W)	B340288a(W)	DB340437!(W)	B340612b(W)	B340780*(W)
B340113*(W)	B340291#(W)	B340444#(W)	DB340614=(W)	B340784=(W)
B340115*(W)	B340304#(W)	B340466#(W)	DB340615+(W)	B340790*(W)
B340126=(W)	B340306*(W)	DB340469 (W)	DB340619#(W)	B340805-(W)
B340128*(W)	DB340311c(W)	DB340475 (W)	B340624#(W)	DB340810b(W)
DB340140!(W)	B340319*(W)	DB340477#(W)	B340632#(W)	DB340811 (W)
B340141=(W)	B340328#(W)	B340482*(W)	B340633-(W)	B340812#(W)
B340150#(W)	DB340330 (W)	B340491#(W)	B340644#(W)	DB340828 (W)
B340170*(W)	B340336#(W)	DB340495 (W)	B340647=(W)	B340831#(W)
B340174*(W)	DB340337 (W)	B340506#(W)	B340659#(W)	DB340832 (W)
DB340190 (W)	B340346*(W)	B340507#(W)	B340669#(W)	B340840"(W)
B340192"(W)	B340352#(W)	B340512#(W)	DB340671#(W)	B340846*(W)
DB340198"(W)	B340368*(W)	B340516#(W)	B340674#(W)	B340852*(W)
B340201-(W)	B340372#(W)	DB340519:	B340685#(W)	DB340859-(W)
B340204#(W)	DB340375c(W)	B340520=(W)	B340692-(W)	B340866 (W)
DB340220 (W)	B340380#(W)	B340529#(W)	B340702-(W)	B340879=(W)
B340225#(W)	B340384a(W)	B340532#(W)	B340711-(W)	B340886*(W)
B340235*(W)	B340387#(W)	DB340536:	DB340712:(W)	DB340895-(W)
B340236#(W)	B340392#(W)	DB340537:	B340716#(W)	DB340902 (W)
B340240#(W)	B340403-(W)	DB340550%(W)		

Number Series: B345000 - B346468

Description: 25t 2 Axle Coal Hopper Wagon
Builder: BR (Shildon Works)
Diagram No.:
Tare Weight: 9.5t
Design Code: HT031F * HT031H + HT031L

Lot No.: 3919
Built: 1977-78
G.L.W.: 31.0t + 31.5t
Tops Code: HTO

B345119*(W)	B345281 (W)	B345382+(W)	B345445 (W)	B345468 (W)
B345202*(W)				

Number Series: B385640 - B387089

Description: 27T 2 Axle Iron Ore Wagon
Builder: BR (Derby Works)
Diagram No.: 1/184
Tare Weight: 9.0t   Fishkind: "BARBEL"
Design Code: ZK006A = ZK006B MS002F + MS002J !

Lot No.: 3089
Built: 1958-59
G.L.W.: 35.5T
Tops Code: MSV +! ZKV

B385642+	DB385664	B385691+(W)	DB385725	DB385748
DB385653	DB385670	DB385697	B385726!	B385749+(W)
DB385654	DB385671	DB385706	DB385734=	B385756+(W)
DB385655	DB385673	DB385709	DB385738 (W)	DB385757
DB385657	DB385675	DB385715	B385739+	DB385759
DB385660	DB385676	DB385716	DB385741=	DB385764
B385661+(W)	B385685+(W)	B385717+(W)	B385742+	DB385767
DB385663	B385686+(W)	DB385722=	B385744+(W)	DB385770

9

DB385777	B385923+	B386101=	DB386270	DB386442
DB385778	B385929+(W)	DB386102=	DB386271	DB386449
DB385781	B385933+(W)	DB386111 (W)	B386273+(W)	DB386453
DB385782	DB385942	B386113+	B386279+(W)	DB386457
B385783+(W)	DB385943	DB386115	DB386282	DB386459
B385786+(W)	B385945+(W)	DB386121	DB386285	DB386463
DB385787	DB385948	DB386132	DB386286=	DB386467=
DB385790	DB385949	DB386134	DB386289	DB386468=
DB385793 (W)	DB385953	DB386138=	DB386291=	DB386469
DB385794	B385954+(W)	DB386140	B386293+(W)	DB386470
B385795+	DB385955=	DB386143	DB386295	DB386476
DB385797=	DB385959	DB386150	DB386300 (W)	DB386479
DB385800	B385961+	DB386155	B386301+	DB386484
DB385801	DB385964	DB386158	B386308+(W)	DB386485
DB385802	DB385967	DB386160	DB386309	DB386495
DB385803	B385970+(W)	DB386162	DB386311	B386498+
DB385807	DB385972	B386165+	DB386317	DB386501
B385809!	DB385979	DB386175	DB386321	DB386503 (W)
DB385810	DB385980	B386176=(W)	DB386324	DB386506
DB385813=	DB385985	B386180+	DB386334	DB386508
DB385814=	DB385990	DB386186	DB386337	B386511+(W)
DB385818	DB386007	B386187+(W)	DB386344	DB386517
DB385820	DB386008	DB386188	B386348+	DB386520
DB385821	DB386009	DB386189=	DB386354 (W)	DB386521
DB385823 (W)	DB386010 (W)	DB386190	DB386355	B386538+(W)
B385824+(W)	DB386019	DB386200	DB386356=	DB386540
DB385825 (W)	DB386021	DB386201	DB386366	DB386546
DB385830 (W)	DB386023	DB386209	DB386367	DB386560=
B385832+(W)	DB386031=	DB386213	B386375+	B386562+(W)
DB385836	DB386032	DB386215	DB386377	DB386570
DB385847	DB386034	B386222+(W)	DB386383	DB386573
DB385859	DB386037 (W)	DB386223	DB386395	DB386575
B385860+(W)	DB386039	B386224+(W)	B386403=	DB386581
B385862+(W)	B386041!	DB386229	DB386404 (W)	B386582+(W)
DB385864	DB386046	DB386230	B386406+	DB386585
B385866+(W)	DB386049 (W)	DB386232	DB386412	B386591+
DB385873	DB386050	DB386237	DB386414	DB386595
DB385876	DB386053	DB386240	B386416+(W)	DB386597
DB385886	DB386056	DB386243	DB386419 (W)	DB386598
DB385887	DB386057 (W)	B386244+	DB386421=	B386603+
DB385888	DB386063 (W)	B386245+	DB386422	B386614+
DB385889	DB386070	DB386246	B386424!(W)	DB386619
B385892+(W)	B386072+(W)	DB386247	B386425!(W)	DB386620
DB385897=	B386076!(W)	DB386249 (W)	DB386426	DB386621
DB385902 (W)	B386081+	DB386251	DB386430	DB386622
DB385903	DB386084	DB386256	B386432+(W)	DB386628
B385905+(W)	B386089+(W)	DB386257	DB386438	DB386635
B385909!(W)	DB386094	DB386258	B386439!(W)	DB386636
DB385910	DB386096=	DB386266	B386441+(W)	DB386637=
DB385919	DB386098			

Number Series: B388090 - B389089

Description: 27T Iron Ore Wagon   Lot No.: 3363
Builder: BR (Derby Works)   Built: 1960-61
Diagram No.: 1/185   G.L.W.: 35.0t
Tare Weight: 8.5t 8.7t +   Fishkind: "BARBEL"
Design Code: ZK007A MS002D +   Tops Code: MSV + ZGV

DB388092	DB388251	DB388409	DB388564	DB388723
B388097+	DB388253	DB388411	DB388567	DB388725
DB388100	DB388254	DB388412	DB388568	B388726+
DB388107	DB388257	B388415+(W)	DB388570	B388728+(W)
DB388108	DB388258 (W)	DB388418	DB388574	B388729+(W)
DB388110	DB388262	DB388419	DB388577	B388730+
DB388112	DB388263	DB388420	DB388578	DB388734
B388118+	DB388272	DB388431	DB388579	DB388735
B388119+(W)	B388273+	DB388433 (W)	DB388588	DB388736
DB388120	B388274+	DB388434	DB388589	DB388738
DB388122 (W)	DB388275	DB388435	DB388592	B388741+(W)
DB388124	DB388276	DB388439 (W)	DB388593	DB388744
DB388126	DB388277	B388443+	DB388596	DB388745
DB388134	DB388278	DB388444	DB388599+(W)	DB388746
DB388137	DB388279	DB388445 (W)	DB388600	DB388749
DB388138	B388288+	DB388447	B388602+(W)	DB388750
B388139+	DB388289	B388448+(W)	DB388604	DB388751
DB388141	DB388295	B388449+(W)	DB388605	DB388755
DB388142	DB388296	DB388451	B388609+(W)	DB388761
DB388145	DB388300	DB388452	DB388611	DB388765
DB388147	DB388303	DB388454	DB388612	DB388766 (W)
DB388148	DB388304	DB388456	DB388613	DB388767
DB388149	DB388305	B388458+	DB388615	DB388769
B388152+	DB388309	DB388462	DB388622	DB388777
B388153+(W)	DB388310	DB388465 (W)	DB388624	DB388781
DB388154	DB388311	B388466+	DB388625	DB388782
DB388157	B388313+	DB388467	DB388632	B388787+(W)
DB388159	DB388314	B388473+	DB388633	DB388788
DB388162	DB388316	DB388475	B388634+	DB388790
DB388163	DB388317	DB388476	DB388635	B388793+(W)
DB388164	B388320+	DB388480	DB388636	DB388796 (W)
B388166+	B388324+(W)	DB388484	DB388639	DB388797
DB388170	DB388325	DB388487	B388644+(W)	B388798+
DB388171	DB388326	DB388490	DB388645	B388799+(W)
DB388175	DB388334	DB388492	DB388647	B388800+(W)
DB388178	DB388336	B388494+	DB388650	DB388804
DB388180	DB388339	DB388495	B388653+	B388807+(W)
B388185+(W)	B388341+	DB388497	DB388654	DB388812+(W)
DB388187	DB388349	DB388499	DB388655	DB388814
DB388188	DB388356	DB388500	DB388656	DB388816
DB388189	DB388361	B388501+(W)	DB388660	DB388817
B388190+(W)	DB388362	DB388502	B388661+	DB388819
DB388191	B388364+(W)	DB388506	DB388662	DB388822
DB388193	DB388368	DB388507	DB388664	DB388826
DB388196	DB388369	B388511+(W)	DB388665	B388827+(W)
DB388198	DB388371	DB388512	DB388667	DB388828
DB388201	DB388374	DB388526	DB388669	DB388830 (W)
DB388205	DB388375	B388531+(W)	DB388671	DB388836
B388209+(W)	DB388377	DB388533	DB388673	DB388838
DB388218	DB388378	DB388534	DB388683	B388845+
DB388219	DB388380	DB388536	DB388686	DB388848
DB388221	DB388382	DB388537	DB388691	DB388849
B388223+(W)	DB388389	DB388543	DB388692	B388850+(W)
B388226+(W)	DB388391	DB388544	B388699+	B388851+(W)
B388230+(W)	DB388392	B388546+	B388704+(W)	DB388854
B388231+(W)	B388398+	DB388548	DB388708 (W)	DB388855 (W)
B388232+	DB388399	DB388551	DB388710	DB388856
DB388244	DB388403	DB388554	B388717+(W)	DB388863
DB388248	DB388406	DB388561	DB388719	DB388865
DB388249	DB388408	B388562+	DB388721	DB388866

DB388868	DB388908	DB388962 (W)	DB388993	DB389046
DB388873	B388916+(W)	DB388963	B388994+(W)	DB389051
DB388874	B388922+(W)	DB388964	DB388996	DB389054
DB388875	DB388923	B388965+	DB388997 (W)	DB389055
DB388877	B388926+	DB388966	DB389000+	B389059+(W)
DB388878	B388931+	DB388967	DB389007+(W)	DB389062
DB388880	DB388932	B388969+(W)	DB389016	DB389064
DB388881	DB388933	DB388970	DB389020	DB389066
DB388882	B388935+(W)	DB388971	DB389023	B389070+
B388886+(W)	DB388936	B388973+(W)	DB389027	DB389075
B388887+	DB388938	DB388976	DB389030	DB389078
DB388893	DB388939	DB388977	DB389033 (W)	DB389079
B388896+	DB388942	DB388982	DB389035	DB389081
DB388899	DB388944	B388984+(W)	DB389039	B389083+(W)
B388901+(W)	DB388947	DB388990	DB389040	DB389084
DB388902	DB388948	DB388991	DB389043	DB389088
DB388906	DB388956	B388992+(W)	DB389045	B389089+
DB388907				

Number Series: B390000 - B390149

Description: 24T 2 Axle Open Sand Wagon
Builder: Standard Wagon Co Ltd   Lot No.: 3859
Diagram No.: 1/193   Built: 1974-75
Tare Weight: 11.0t   Fishkind: "ZANDER" G.L.W.: 35.5t
Design Code: ZK008A   Tops Code: ZKV

DB390000	DB390034	DB390064	DB390093	DB390122
DB390002	DB390035	DB390065	DB390094	DB390123
DB390003	DB390036	DB390066	DB390095	DB390124
DB390004	DB390037	DB390067	DB390096	DB390125
DB390005	DB390038	DB390068	DB390097	DB390126
DB390006	DB390039	DB390069	DB390098	DB390127
DB390007	DB390040	DB390070	DB390099	DB390128
DB390008	DB390041	DB390071	DB390100	DB390129
DB390009	DB390042	DB390072	DB390101	DB390130
DB390010	DB390043	DB390073	DB390102	DB390131
DB390011	DB390044	DB390074	DB390103 (W)	DB390132
DB390012	DB390045	DB390075	DB390104	DB390133
DB390013	DB390046	DB390076	DB390105	DB390134
DB390014	DB390047	DB390077	DB390106	DB390135
DB390015	DB390048	DB390078	DB390107	DB390136
DB390016	DB390049	DB390079	DB390108	DB390137
DB390017	DB390051 (W)	DB390080	DB390109	DB390138
DB390019	DB390052	DB390081	DB390110	DB390139
DB390020	DB390053	DB390082	DB390111	DB390140
DB390021	DB390054	DB390083	DB390112	DB390141
DB390023	DB390055	DB390084	DB390113	DB390142
DB390024	DB390056	DB390085	DB390114	DB390143
DB390026 (W)	DB390057	DB390086	DB390115	DB390144
DB390028	DB390058	DB390087	DB390116	DB390145
DB390029	DB390059	DB390088 (W)	DB390117	DB390146
DB390030	DB390060	DB390089	DB390119	DB390147
DB390031	DB390061	DB390090	DB390120	DB390148
DB390032	DB390062	DB390091	DB390121	DB390149
DB390033	DB390063	DB390092		

12

Number Series: B413950 - B414649

Description: 21T 2 Axle Coal Hopper Wagon
Builder: BR (Shildon Works)   Lot No.: 2330
Diagram No.: 1/146   Built: 1952
Tare Weight: 10.5t   G.L.W.: 32.0t
Design Code: ZD152B   Tops Code: ZDV

  DB413963 (W)

Number Series: B415150 - B415949

Description: 21T 2 Axle Coal Hopper Wagon
Builder: BR (Shildon Works)   Lot No.: 2552
Diagram No.: 1/146   Built: 1954
Tare Weight: 9.45t + 10.5t   G.L.W.: 32.0t
Design Code: HT018C + ZD152A * ZD152B   Tops Code: HTV + ZDV

  B415409+(W)   DB415598*(W)   DB415891 (W)

Number Series: B415950 - B416449

Description: 21T 2 Axle Coal Hopper Wagon
Builder: BR (Shildon Works)   Lot No.: 2713
Diagram No.: 1/146   Built: 1955
Tare Weight: 9.45t * 10.5t   G.L.W.: 32.0t
Design Code: HT018C * ZD123E   Tops Code: HTV * ZDV

  B416187*(W)   B416199*(W)   DB416438 (W)

Number Series: B416450 - B416749

Description: 21T 2 Axle Coal Hopper Wagon
Builder: Cravens C & W Co Ltd   Lot No.: 2726
Diagram No.: 1/146   Built: 1955
Tare Weight: 9.45t * 10.5t   G.L.W.: 32.0t
Design Code: HT018C * ZD152B   Tops Code: HTV * ZDV

  DB416457*(W)   B416500*(W)   DB416527 (W)   B416660*(W)

Number Series: B417550 - B419199

Description: 21T 2 Axle Coal Hopper Wagon
Builder: BR (Shildon Works)   Lot No.: 2731
Diagram No.: 1/146   Built: 1955
Tare Weight: 10.5t   G.L.W.: 32.0t
Design Code: ZD152A   Tops Code: ZDV

  DB418096 (W)   DB418249 (W)   DB419140 (W)   DB419188 (W)

Number Series: B419200 - B419249

Description: 21T 2 Axle Coal Hopper Wagon
Builder: BR (Shildon Works)   Lot No.: 2954
Diagram No.: 1/146   Built: 1956
Tare Weight: 10.5t   G.L.W.: 32.0t
Design Code: ZD152A   Tops Code: ZHV

  DB419221 (W)

Number Series: B419250 - B419999

Description: 21T 2 Axle Coal Hopper Wagon
Builder: Birmingham R C & W Co Ltd   Lot No.: 2932
Diagram No.: 1/146   Built: 1956
Tare Weight: 9.45t * 10.5t   G.L.W.: 32.0t
Design Codes: HT018C * ZD152A   Tops Code: HTV * ZDV

  B419469*(W)   B419721*(W)   DB419748   B419852*(W)   B419957*(W)
  DB419547 (W)   B419738*(W)   B419787*(W)

Number Series: B420000 - B420699

Description: 21T 2 Axle Coal Hopper Wagon   Lot No.: 2933
Builder: Cravens R C & W Co Ltd   Built: 1956
Diagram No.: 1/146   G.L.W.: 32.0t
Tare Weight: 9.45t #* 10.5t   Tops Code: HTV #* ZDV
Design Code: HT018C * HT018D # ZD123E - ZD152B

  DB420093-(W)   DB420247 (W)   DB420294   LDB420360 (W)   DB420494 (W)
  DB420142 (W)   B420250#(W)   DB420323*(W) DB420438-(W)   B420536#(W)
  DB420159 (W)   B420286#(W)

Number Series: B420700 - B421249

Description: 21T 2 Axle Coal Hopper Wagon   Lot No.: 2934
Builder: Fairfields S & E Co Ltd   Built: 1956
Diagram No.: 1/146   G.L.W.: 32.0t
Tare Weight: 9.45t * 10.5t   Tops Code: HTV * ZDV
Design Code: HT018C * ZD123G = ZD152A + ZD152B

  DB420733+(W)   B420796*(W)   DB420972*   DB421081 (W)   DB421110=(W)
  DB420756 (W)   DB420930 (W)   DB421018*(W) DB421103+(W)   DB421123
  DB420791 (W)

Number Series: B421250 - B422249

Description: 21T 2 Axle Coal Hopper Wagon
Builder: Metropolitan Cammell Co Ltd   Lot No.: 2935
Diagram No.: 1/146   Built: 1956
Tare Weight: 9.45t * 10.5t   G.L.W.: 32.0t
Design Code: HT018C * ZD0152A ZD152B +   Tops Code: HTV * ZDV

14

```
DB421462+(W)   DB421699*(W)   DB422078*(W)   DB422167*(W)   DB422229*(W)
DB421576 (W)   DB421810*(W)   DB422082 (W)   DB422196 (W)
```

Number Series: B422250 - B422749

Description: 21T 2 Axle Coal Hopper Wagon
Builder: Fairfields S & E Co Ltd   Lot No.: 3013
Diagram No.: 1/146   Built: 1957
Tare Weight: 9.45t *# 10.5t   G.L.W.: 32.0t
Design Code: HT018C * HT018D # ZD152B   Tops Code: HTV #* ZDV

```
DB422263 (W)   DB422515 (W)   DB422608 (W)   B422682#(W)   DB422733 (W)
DB422307*(W)   DB422570       B422639#(W)   DB422689#(W)   DB422747 (W)
 B422332*(W)   DB422575#(W)   B422647#(W)   DB422719 (W)   DB422748 (W)
DB422344*(W)   DB422598 (W)   B422681#(W)
```

Number Series: B423050 - B423549

Description: 21T 2 Axle Coal Hopper Wagon
Builder: Birmingham R C & W Co Ltd   Lot No.: 3030
Diagram No.: 1/146   Built: 1957
Tare Weight: 9.45t * 10.5t   G.L.W.: 32.0t
Design Code: HT018D * ZD152B   Tops Code: HTV * ZDV

```
 B423104*(W)    B423150*(W)   DB423211 (W)   DB423326 (W)   DB423432 (W)
DB423105 (W)    B423152*(W)   DB423215 (W)   DB423345 (W)    B423443*(W)
DB423111 (W)   DB423155 (W)    B423250*(W)   DB423351 (W)   DB423496 (W)
DB423128 (W)    B423158*(W)   DB423262 (W)   DB423394 (W)   DB423504 (W)
DB423129 (W)   DB423181 (W)    B423265*(W)   DB423407 (W)    B423543*(W)
 B423147*(W)   DB423189 (W)   DB423302 (W)
```

Number Series: B423550 - B424549

Description: 21T 2 Axle Coal Hopper Wagon
Builder: Cravens R C & W Co Ltd   Lot No.: 3031
Diagram No.: 1/146   Built: 1957
Tare Weight: 10.5t   G.L.W.: 32.0t
Design Code: HT018D # ZD151A + ZD152B   Tops Code: HTV # ZDV

```
DB423562 (W)    B423720#(W)   DB423846 (W)    B424324#(W)   DB424410 (W)
 B423569#(W)    B423723#(W)   DB423849 (W)   DB424328 (W)   DB424434
DB423594 (W)   DB432732 (W)   DB423853 (W)   DB424331 (W)   DB424444 (W)
DB423639 (W)    B423753#(W)   DB423871 (W)   DB424333 (W)   DB424457 (W)
 B423657#(W)   DB423760       DB423952 (W)   DB424337 (W)   DB424503 (W)
DB423668 (W)   DB423765 (W)    B424218#(W)   DB424351 (W)   DB424538 (W)
DB423690 (W)   DB423771 (W)   DB424231 (W)   DB424359 (W)   DB424540 (W)
DB423703 (W)    B423804#(W)    B424260#(W)   DB424393 (W)    B424541#(W)
DB423719 (W)   DB423806 (W)   DB424313 (W)   DB424403
```

Number Series: B424550 - B425349

Description: 21T 2 Axle Coal Hopper Wagon
Builder: Gloucester R C & W Co Ltd   Lot No.: 3032
Diagram No.: 1/146   Built: 1957
Tare Weight: 10.5t   G.L.W.: 32.0t
Design Code: HT018D # ZD152B   Tops Code: HTV # ZDV

```
DB424607 (W)    DB424798 (W)    DB425052 (W)    B425120#(W)    DB425215
DB424670 (W)    DB424817 (W)    DB425056 (W)    DB425127 (W)    B425217#(W)
DB424687 (W)    DB424833 (W)    B425077#(W)    DB425130 (W)    DB425219 (W)
DB424702 (W)    DB424865 (W)    DB425096 (W)    DB425144 (W)    DB425222 (W)
DB424734 (W)    DB424905 (W)    DB425100 (W)    DB425211 (W)    DB425232 (W)
DB424773 (W)    B424932#(W)    DB425117 (W)    DB425212 (W)    B425313#(W)
 B424793#(W)    DB424983 (W)
```


Number Series: B425350 - B425849

Description: 21T 2 Axle Coal Hopper Wagon
Builder: Head Wrightson & Co Ltd   Lot No.: 3033
Diagram No.: 1/146   Built: 1957
Tare Weight: 10.05t   G.L.W.: 32.0t
Design Code: ZD152B   Tops Code: ZDV

```
DB425352 (W)    DB425472 (W)    DB425520 (W)    DB425562 (W)    DB425674 (W)
DB425422 (W)    DB425475 (W)    DB425532 (W)    DB425612 (W)    DB425730 (W)
DB425437 (W)    DB425492 (W)    DB425541 (W)    DB425654 (W)    DB425822 (W)
DB425467 (W)    DB425510 (W)    DB425561 (W)    DB425655 (W)    LDB425840 (W)
```


Number Series: B425850 - B426149

Description: 21T 2 Axle Coal Hopper Wagon
Builder: Head Wrightson & Co Ltd   Lot No.: 3034
Diagram No.: 1/146   Built: 1957
Tare Weight: 10.5t   G.L.W.: 32.0t
Design Code: ZD152B   Tops Code: ZDV

```
DB425923 (W)    DB425971 (W)    DB426011        DB426112 (W)    DB426142 (W)
DB425935 (W)    DB425994 (W)    DB426036 (W)    DB426116 (W)    DB426144 (W)
DB425946        DB425996 (W)    DB426063 (W)    DB426127 (W)
```


Number Series: B426150 - B426949

Description: 21T 2 Axle Coal Hopper Wagon
Builder: Hurst Nelson & Co Ltd   Lot No.: 3035
Diagram No: 1/146   Built: 1957
Tare Weight: 10.5t   Fishkind: "TOPE" + G.L.W.: 32.0t
Design Code: HT018D # ZD151A + ZD152B   Tops Code: HTV # ZDV +

```
 B426159#(W)    B426276#(W)    DB426376 (W)    B426538#(W)    DB426672 (W)
DB426163 (W)    DB426289 (W)    B426397#(W)    DB426575 (W)    DB426680 (W)
 B426165#(W)    DB426317 (W)    DB426407 (W)    DB426616 (W)    B426688#(W)
DB426194+       DB426323 (W)    DB426447 (W)    DB426618 (W)    DB426724 (W)
 B426225#(W)    DB426368 (W)    DB426459 (W)    DB426624        DB426750 (W)
```

```
DB426756 (W)    DB426773 (W)    DB426841 (W)    B426899#(W)    DB426927 (W)
DB426759 (W)    DB426785 (W)    B426862#(W)    DB426917 (W)    DB426929 (W)
DB426767 (W)    DB426813 (W)    DB426891 (W)    DB426922 (W)
```

Number Series: B426950 - B427399

Description: 21T 2 Axle Coal Hopper Wagon
Builder: Metropolitan Cammell Co Ltd   Lot No.: 3036
Diagram No.: 1/146   Built: 1957
Tare Weight: 10.5t   G.L.W.: 32.0t
Design Codes: HT018D #   ZD152B   Tops Code: HTV #   ZDV

```
DB426957 (W)    DB427122 (W)    DB427219 (W)    DB427309 (W)    DB427343 (W)
DB426970 (W)    DB427137 (W)    DB427229 (W)    DB427313 (W)    B427346#(W)
DB427027 (W)    DB427168 (W)    DB427245 (W)    DB427318 (W)    DB427349 (W)
DB427086 (W)    B427173#(W)    DB427247 (W)    DB427319 (W)    DB427357 (W)
DB427095 (W)    DB427194 (W)    DB427274 (W)    DB427338 (W)    B427376#(W)
DB427102 (W)    DB427212 (W)    DB427283 (W)
```

Number Series: B427400 - B428549

Description: 21T 2 Axle Coal Hopper Wagon
Builder: Charles Roberts Co Ltd   Lot No.: 3037
Diagram No.: 1/146   Built: 1957
Tare Weight: 10.5t   G.L.W.: 32.0t
Design Code: HT018D #   ZD152B   Tops Code: HTV #   ZDV

```
DB427401 (W)    DB427622 (W)    B427920#(W)    DB428032 (W)    DB428323 (W)
DB427415 (W)    DB427634 (W)    B427930#(W)    DB428039 (W)    DB428325
DB427436 (W)    DB427678 (W)    DB427959 (W)    DB428051        DB428342 (W)
DB427456 (W)    DB427709 (W)    DB427960 (W)    B428128#(W)    DB428421 (W)
DB417474 (W)    DB427725 (W)    DB427965 (W)    B428204#(W)    DB428441 (W)
DB427491 (W)    B427727#(W)    DB427968 (W)    DB428207 (W)    DB428476 (W)
DB427509 (W)    DB427818 (W)    DB427988 (W)    DB428241 (W)    B428500#(W)
B427578#(W)    DB427820 (W)    DB428005 (W)    DB428258 (W)    DB428531 (W)
DB427583 (W)    DB427864 (W)    DB428009        DB428317 (W)    B428537#(W)
DB427611 (W)    DB427869 (W)    DB428019 (W)    DB428318 (W)
```

Number Series: B428550 - B429049

Description: 21T 2 Axle Coal Hopper Wagon   Lot No.: 3045
Builder: Standard R C & W Co Ltd   Built: 1957
Diagram No.: 1/146   G.L.W.: 32.0t
Tare Weight: 10.5t   Tops Code: HTO #= ZDV +
Design Code: HT018C #   HT018D =   ZD152A +   ZD152B

```
B428598#(W)    DB428655+(W)    B428928#(W)    DB428959 (W)    DB429025 (W)
DB428616 (W)    DB428757 (W)    DB428939 (W)    B428982=(W)    B429049=(W)
B428650=(W)    DB428845 (W)    B428941#(W)
```

Number Series: B429350 - B429499

Description: 21T 2 Axle Coal Hopper Wagon   Lot No.: 3158
Builder: Birmingham R C & W Co Ltd   Built: 1958
Diagram No.: 1/146   G.L.W.: 32.0t
Tare Weight: 9.45t += 10.5t   Tops Code: HTO + HTV =
Design Code: HT004B + HT018D = ZD152B   ZDV

  B429357=(W)   DB429424   DB429455 (W)   B429466=(W)   B429467=(W)
DB429359 (W)   B429433+(W)

Number Series: B429500 - B429649

Description: 21T 2 Axle Coal Hopper Wagon
Builder: Gloucester R C & W Co Ltd   Lot No.: 3159
Diagram No.: 1/146   Built: 1959
Tare Weight: 9.45t = 10.5t   G.L.W.: 32.0t
Design Code: HT018D = ZD152B   Tops Code: HTV = ZDV

  DB429546=(W)   DB429588 (W)   DB429608=(W)   DB429609 (W)   DB429630 (W)

Number Series: B429650 - B429799

Description: 21T 2 Axle Coal Hopper Wagon
Builder: Hurst Nelson & Co Ltd   Lot No.: 3160
Diagram No.: 1/146   Built: 1958
Tare Weight: 9.45t # 10.45t* 10.5t   G.L.W.: 32.0t
Design Code: HT018D # ZD123E * ZD152B   Tops Code: HTV # ZDV

  DB429653 (W)   DB429670 (W)   DB429742*(W)   DB429783 (W)   DB429798 (W)
  DB429659   B429675#(W)   B429765#(W)   DB429784 (W)   DB429799*(W)
  DB429665*(W)   DB429678*(W)

Number Series: B429800 - B430799

Description: 21T 2 Axle Coal Hopper Wagon   Lot No.: 3120
Builder: BR (Shildon Works)   Built: 1958
Diagram No.: 1/149   G.L.W.: 32.0t
Tare Weight: 9.55t + 9.8t = 10.5t   Tops Code: HTV =+ ZDV
Design Code: HT019B = HT021B + ZD152C ZD153A # ZD153B *

  DB429810+(W)   DB430097 (W)   DB430264 (W)   DB430438 (W)   DB430653*(W)
  DB429908*(W)   B430110+(W)   B430275=(W)   B430448=(W)   DB430659 (W)
  DB429950 (W)   DB430138 (W)   DB430316#(W)   DB430453*(W)   B430674=(W)
  DB429958 (W)   DB430139 (W)   DB430323 (W)   DB430455+(W)   LDB430679*
  DB430014 (W)   DB430140*(W)   DB430328*(W)   LDB430462 (W)   DB430722 (W)
  DB430022+(W)   B430141=(W)   DB430357 (W)   DB430515 (W)   DB430725 (W)
  B430025+(W)   DB430157 (W)   DB430364*(W)   DB430589 (W)   B430731+(W)
  B430056=(W)   DB430221 (W)   B430375=(W)   DB430610 (W)   DB430789#
  B430077=(W)   DB430226 (W)   B430399=(W)   DB430623#(W)   DB430798 (W)
LDB430080 (W)   DB430228 (W)   DB430421 (W)

Number Series: B430800 - B433749

Description: 21T 2 Axle Coal Hopper Wagon   Lot No.: 3157
Builder: Pressed Steel Co Ltd   Built: 1958
Diagram No.: 1/146   G.L.W.: 31.0t + 32.0t
Tare Weight: 9.45t #+ 10.45t * 10.5t   Tops Code: HTV +# ZDV
Design Code: HT004D + HT018D # ZD123E * ZD152B ZD152C =

B430825#(W)	DB431510 (W)	B432042#(W)	B432642#(W)	B433217#(W)
B430857#(W)	B431522#(W)	DB432126 (W)	DB432643 (W)	B433220#(W)
DB430869 (W)	B431535#(W)	DB432129*(W)	B432648#(W)	B433246#(W)
DB430872 (W)	B431555+(W)	DB432141 (W)	B432667+(W)	B433280#(W)
B430936#(W)	DB431589 (W)	DB432167 (W)	DB432679 (W)	B433286#(W)
B430942+(W)	DB431611 (W)	DB432173*(W)	B432686#(W)	B433298#(W)
DB430972	B431614#(W)	B432180+(W)	B432701#(W)	DB433301 (W)
DB431042 (W)	DB431619 (W)	DB432181 (W)	DB432729 (W)	DB433303 (W)
B431048+(W)	DB431624*(W)	DB432218 (W)	DB432734 (W)	B433307#(W)
B431088#(W)	B431631#(W)	B432233#(W)	B432740#(W)	DB433323 (W)
B431098+(W)	DB431653 (W)	DB432240 (W)	B432772#(W)	B433353#(W)
DB431111*(W)	DB431702*(W)	B432243#(W)	B432818#(W)	B433368#(W)
B431116#(W)	B431712+(W)	DB432303=(W)	B432825+(W)	B433370#(W)
B431160#(W)	B431724+(W)	B432346#(W)	DB432879 (W)	DB433374*(W)
B431161+(W)	B431746#(W)	B432356#(W)	B432889#(W)	DB433375*(W)
DB431180 (W)	DB431752 (W)	DB432363 (W)	B432905#(W)	B433383#(W)
DB431218 (W)	DB431767 (W)	DB432389*(W)	DB432921 (W)	B433447#(W)
B431232#(W)	DB431774 (W)	DB432392 (W)	DB432958*(W)	B433460#(W)
DB431248 (W)	B431810#(W)	DB432406+(W)	B432964#(W)	B433461#(W)
DB431251 (W)	DB431835 (W)	DB432415 (W)	DB432972 (W)	DB433466 (W)
B431274#(W)	DB431836 (W)	DB432433 (W)	DB432973 (W)	B433476#(W)
B431278#(W)	DB431913*(W)	DB432466 (W)	B433009#(W)	DB433479 (W)
B431312#(W)	DB431918 (W)	DB432481 (W)	DB433010 (W)	DB433511 (W)
B431333#(W)	DB431939 (W)	B432493#(W)	DB433049 (W)	DB433557 (W)
B431350#(W)	DB431974 (W)	DB432529 (W)	B433079#(W)	B433633+(W)
B431417#(W)	DB431976 (W)	DB432544 (W)	DB433129 (W)	DB433702
B431465#(W)	DB432000 (W)	B432559#(W)	B433213#(W)	DB433712 (W)
DB431481 (W)	B432024#(W)	B432620#(W)	DB433216 (W)	B433716#(W)
B431498#(W)	DB432027 (W)			

Number Series: B437500 - B437899

Description: 25.5T 2 Axle Ironstone Hopper Wagon
Builder: BR (Shildon Works)   Lot No.: 3002
Diagram No.: 1/167   Built: 1957
Tare Weight: 11.5t 11.55t =*+   G.L.W.: 45.5t
Design Code: HJ004A = HK003B   Tops Code: HJV = HKV
             ZD140A * ZD141A +   ZDV *+

B437513 (W)	B437569 (W)	B437637 (W)	B437683 (W)	B437697 (W)
DB437519+(W)	B437621 (W)	B437653=(W)	B437685=(W)	DB437781+(W)
B437523=(W)	B437623=(W)	B437682=(W)	B437696=(W)	DB437859+(W)
DB437535*(W)	B437665 (W)			

Number Series: B438900 - B439499

Description: 25.5T 2 Axle Ironstone Hopper Wagon
Builder: BR (Shildon Works)   Lot No.: 3001
Diagram No.: 1/163   Built: 1957
Tare Weight: 11.1t   G.L.W.: 35.5t
Design Code: HJ002D   Tops Code: HJV

  B439496 (W)

Number Series: B439500 - B439699

Description: 25.5T 2 Axle Ironstone Hopper Wagon
Builder: BR (Shildon Works)   Lot No.: 3142
Diagram No.: 1/165   Built: 1957
Tare Weight: 10.5t   G.L.W.: 44.0t
Design Code: HJ002A   Tops Code: HJV

B439521 (W)	B439561 (W)	B439597 (W)	B439621 (W)	B439646 (W)
B439523 (W)	B439590 (W)	B439617 (W)	B439623 (W)	B439685 (W)
B439527 (W)	B439595 (W)	B439620 (W)	B439624 (W)	B439691 (W)
B439554 (W)				

Number Series: B439700 - B440049

Description: 25.5T 2 Axle Ironstone Hopper Wagon
Builder: BR (Shildon Works)   Lot No.: 3189
Diagram No.: 1/166   Built: 1959
Tare Weight: 9.45t   G.L.W.: 34.0t
Design Code: HJ002B   Tops Code: HJV

B439733 (W)	B439768 (W)	B439792 (W)	B439806 (W)	B439824 (W)
B439750 (W)	B439775 (W)	B439793 (W)	B439809 (W)	B439830 (W)
B439754 (W)	B439776 (W)	B439795 (W)	B439812 (W)	B439838 (W)
B439756 (W)	B439779 (W)	B439796 (W)	B439816 (W)	B439869 (W)
B439761 (W)	B439788 (W)	B439798 (W)	B439818 (W)	B439879 (W)
B439765 (W)	B439789 (W)	B439803 (W)	B439822 (W)	B439948 (W)

Number Series: B450400 - B451399

Description: 13T Low Goods Wagon
Builder: BR (Shildon Works)   Lot No.: 2194
Diagram No.: 1/002   Built: 1951
Tare Weight: 6.0t # 7.0t   G.L.W.: 19.0t # 20.0t
Design Code: ZD069A # ZV078A + ZV091A   Tops Code: ZDV # ZVV

ADB450460+   KDB450590 (W) KDB450699 (W) DB450957#(W) KDB451159 (W)

Number Series: B451700 - B451899

Description: 13T Low Goods Wagon
Builder: BR (Shildon Works)
Diagram No.: 1/002
Tare Weight: 7.0t
Design Code: ZV091A

Lot No.: 2340
Built: 1952
G.L.W.: 20.0t
Tops Code: ZVV

KDB451884 (W)

Number Series: B451900 - B452199

Description: 13T Low Goods Wagon
Builder: BR (Shildon Works)
Diagram No.: 1/002
Tare Weight: 7.0t
Design Code: ZV091A

Lot No.: 2420
Built: 1953
G.L.W.: 20.0t
Tops Code: ZVV

KDB452068 (W) KDB452109 (W)

Number Series: B452200 - B452399

Description: 13T Low Goods Wagon
Builder: BR (Shildon Works)
Diagram No.: 1/002
Tare Weight: 6.0t * 7.0t 9.0t +
Design Code: ZD069D * ZS002C + ZV091A
Tops Code: ZDP * ZSW + ZVV

Lot No.: 2461
Built: 1953
G.L.W.: 9.0t + 18.0t *
          20.0t

DB452219*(W) KDB452240 (W) ADB452339+   KDB452374 (W)

Number Series: B452400 - B452599

Description: 13T Low Goods Wagon
Builder: BR (Shildon Works)
Diagram No.: 1/002
Tare Weight: 7.0t
Design Code: ZV091A

Lot No.: 2467
Built: 1955
G.L.W.: 20.0t
Tops Code: ZVV

KDB452479 (W)

Number Series: B452600 - B452899

Description: 13T Low Goods Wagon
Builder: BR (Shildon Works)
Diagram No.: 1/002
Tare Weight: 7.0t 13.0t +
Design Code: ZS002C + ZV091A

Lot No.: 2729
Built: 1957
G.L.W.: 13.0t + 20.0t
Tops Code: ZSW + ZVV

ADB452604+   KDB452611 (W) KDB452615 (W) ADB452732+

21

Number Series: B452900 - B453499

Description: 13T Low Goods Wagon
Builder: BR (Shildon Works)
Diagram No.: 1/002
Tare Weight: 6.0t =*- 7.0t 13.0t +
Design Code: ZD069A = ZD069D * ZS002C + ZV091A ZY084C -
Tops Code: ZDO = ZDP * ZSW + ZVV ZYW -

Lot No.: 2998
Built: 1959
G.L.W.: 13.0t + 20.0t
          19.0t =*-

```
LDB453037-    KDB453133 (W)  DB453255*    LDB453305-    KDB453322 (W)
KDB453084=(W) KDB453176=(W) KDB453280 (W) LDB453315-    DB453343=(W)
KDB453126=(W) ADB453241+
```

Number Series: B457200 - B457596

Description: 13T Medium Goods Wagon
Builder: BR (Wolverton Works)
Diagram No.: 1/017
Tare Weight: 6.1t + 6.6t
Design Code: ZA002A + ZD001B

Lot No.: 2108
Built: 1951
G.L.W.: 19.0t
Tops Code: ZAV + ZDV

```
DB457376+(W) ADB457543 (W)
```

Number Series: B457597 - B458596

Description: 13T Medium Goods Wagon
Builder: BR (Ashford Works)
Diagram No.: 1/019
Tare Weight: 6.5t * 7.0t + 7.1t
Design Code: ZA004B   ZD002B #
             ZV203A * ZX064E +

Lot No.: 2235
Built: 1951
G.L.W.: 19.5t * 20.0t
Tops Code: ZAV   ZDV #
           ZVV * ZXV +

```
DB457615+(W)  DB458079 (W)  DB458267 (W)  DB458393 (W)  DB458515*(W)
DB458038*(W)  DB458250 (W)  DB458376 (W) ADB458484#(W)
```

Number Series: B458597 - B459596

Description: 13T Medium Goods Wagon
Builder: BR (Ashford Works)
Diagram No.: 1/019
Tare Weight: 6.5t 7.0t * 7.1t -+
Design Code: ZA004A   ZD002B -
             ZR009A + ZV047A *

Lot No.: 2236
Built: 1951
G.L.W.: 19.5t 20.0t *+
Tops Code: ZDV - ZRV +
           ZVV

```
ADB458608*(W)  DB458783 (W)  DB459040 (W)  DB459309 (W) ADB459394+(W)
 DB458756 (W)  DB458813+(W) KDB459193-(W)
```

Number Series: B459597 - B460596

Description: 13T Medium Goods Wagon
Builder: BR (Ashford Works)   Lot No.: 2351 and 2430
Diagram No.: 1/019   Built: 1952
Tare Weight: 6.5t #*=+ 7.1t   G.L.W.: 19.5t #* 20.0t
Design Code: ZA004A #   ZA004B   Tops Code: ZAO # ZAV
             ZD002B +   ZV203A *   ZDV + ZVV *

 DB459707*(W) KDB460178+(W) DB460247*(W) DB460266=(W) DB460406 (W)
 DB459994 (W) DB460215 (W)   DB460263#   DB460293*(W) DB460506 (W)
 DB460070*(W) DB460239 (W)

Number Series: B460997 - B461596

Description: 13T Medium Goods Wagon   Lot No.:2488
Builder: BR (Ashford Works)   Built: 1952
Diagram No.: 1/019   G.L.W.: 19.5t 20.0t *+
Tare Weight: 6.5t 7.1t *+   Tops Code: ZAV * ZDV +
Design Code: ZA004B * ZD002B + ZV203A   ZVV

 DB461051*(W) DB461139 (W)   DB461230+(W) DB461457*(W) DB461551*(W)
 DB461120*(W) DB461225*(W)   DB461438 (W) DB461481*(W)

Number Series: B462707 - B462796

Description: Adaptor Wagon (formerly 16T Palbrick Wagon)
Builder: BR (Ashford Works)   Lot No.: 3243
Diagram No.: 1/087   Built: 1959
Tare Weight: 8.85t + 9.0t   G.L.W.: 9.0t
Design Code: RF001A   RG001A +   Tops Code: RFQ   RGQ +

ADB462709 (W) ADB462733   ADB462755+(W) ADB462762+(W) ADB462775
ADB462712   ADB462734 (W)   ADB462756+   ADB462763+   ADB462778+
ADB462713+(W) ADB462735 (W)   ADB462758+   ADB462769   ADB462779+
ADB462721+(W) ADB462748+   ADB462759+   ADB462771   ADB462783+
ADB462728+   ADB462749   ADB462760+   ADB462772   ADB462784 (W)
ADB462730 (W) ADB462751 (W)   ADB462761+(W) ADB462773+   ADB462788+
ADB462732+   ADB462753

Number Series: B475050 - B477049

Description: 13T High Goods Wagon
Builder: BR (Shildon Works)   Lot No.: 2128
Diagram No.: 1/019   Built: 1950
Tare Weight: 7.4t 8.1t +   G.L.W.: 20.5t 21.0t +
Design Code: ZG011A ZG083A +   Tops Code: ZGB +   ZGV

ADB476462 (W) ADB476466+(W)

Number Series: B477050 - B477649

Description: 13T High Goods Wagon
Builder: BR (Ashford Works)
Diagram No.: 1/034
Tare Weight: 7.0t
Design Code: ZG008B

Lot No.: 2153
Built: 1950
G.L.W.: 20.0t
Tops Code: ZGV

B477209

Number Series: B477650 - B479149

Description: 13T High Goods Wagon
Builder: BR (Derby Works)
Diagram No.: 1/039
Tare Weight: 6.5t
Design Code: ZY113B

Lot No.: 2179
Built: 1951
G.L.W.: 20.5t
Tops Code: ZYW

LDB477799

Number Series: B479150 - B480649

Description: 13T High Goods Wagon
Builder: BR (Shildon Works)
Diagram No.: 1/041
Tare Weight: 7.4t
Design Code: ZG015D

Lot No.: 2195
Built: 1951
G.L.W.: 20.3t
Tops Code: ZGV

ADB479731 (W)

Number Series: B480650 - B482149

Description: 13T High Goods Wagon
Builder: BR (Shildon Works)
Diagram No.: 1/041
Tare Weight: 8.1t
Design Code: ZG083A

Lot No.: 2196
Built: 1951
G.L.W.: 21.0t
Tops Code: ZGB

LDB481682 (W) ADB481866 (W)

Number Series: B482150 - B483649

Description: 13T High Goods Wagon
Builder: BR (Ashford Works)
Diagram No.: 1/041
Tare Weight: 8.1t
Design Code: ZG083A

Lot No.: 2197
Built: 1951
G.L.W.: 21.0t
Tops Code: ZGB

B482804 (W)

Number Series: B483650 - B483749

Description: 13T High Goods Wagon
Builder: BR (Ashford Works)   Lot No.: 2061
Diagram No.: 1/016   Built: 1949
Tare Weight: 6.0t   G.L.W.: 19.0t
Design Code: ZV080A   Tops Code: ZVO

LDB483707 (W)

Number Series: B484150 - B484199

Description: 12T Pipe Wagon
Builder: BR (Wolverton Works)   Lot No.: 3070
Diagram No.: 1/462   Built: 1956
Tare Weight: 8.4t   G.L.W.: 8.4t #   20.5t
Design Code: ZD108A   ZD108E + ZD108F * Tops Code: ZDV   ZDW +*
             ZX167B #   ZXV #

KDB484151*   KDB484159 (W) KDB484173 (W) KDB484191*   KDB484198*
 DB484154 (W) KDB484163+   KDB484176 (W) KDB484196#   KDB484199
 DB484158*(W) KDB484164*   KDB484177+

Number Series: B486000 - B486749

Description: 13T High Goods Wagon
Builder: BR (Derby Works)   Lot No.: 2462
Diagram No.: 1/039   Built: 1953
Tare Weight: 7.4t 7.5t +   G.L.W.: 20.5t
Design Code: ZG013B + ZY113A   Tops Code: ZGV +   ZYV

LDB486055   ADB486121+(W)

Number Series: B486750 - B487719

Description: 13T High Goods Wagon
Builder: BR (Shildon Works)   Lot No.: 2341
Diagram No.: 1/146   Built: 1952
Tare Weight: 7.4t   G.L.W.: 20.3t
Design Code: ZG015D   Tops Code: ZGV

 DB486945 (W)

Number Series: B487720 - B488749

Description: 13T High Goods Wagon
Builder: BR (Shildon Works)   Lot No.: 2342
Diagram No.: 1/047   Built: 1953
Tare Weight: 7.4t   G.L.W.: 20.5t
Design Code: ZG020B   Tops Code: ZGV

ADB487982 (W)

25

Number Series: B488750 - B490249

Description: 13T High Goods Wagon
Builder: Birmingham R C & W Co Ltd
Diagram No.: 1/041
Tare Weight: 7.4t + 8.1t
Design Code: ZG015D + ZG083A

Lot No.: 2361
Built: 1953
G.L.W.: 20.3t + 21.0t
Tops Code: ZGB   ZGV +

ADB488985   ADB489405   KDB490103+(W)

Number Series: B491700 - B492199

Description: 13T High Goods Wagon
Builder: BR (Ashford Works)
Diagram No.: 1/039
Tare Weight: 7.5t
Design Code: ZG013B

Lot No.: 2723
Built: 1955
G.L.W.: 20.5t
Tops Code: ZGV

ADB492135 (W)

Number Series: B492700 - B493899

Description: 13T High Goods Wagon
Builder: Gloucester R C & W Co Ltd
Diagram No.: 1/044
Tare Weight: 6.5t
Design Code: ZY109B

Lot No.: 2396
Built: 1955
G.L.W.: 19.5t
Tops Code: ZYW

LDB493293   LDB493744

Number Series: B493900 - B494769

Description: 13T High Goods Wagon
Builder: R. Y. Pickering & Co Ltd
Diagram No.: 1/044
Tare Weight: 6.0t 6.5t +
Design Codes: ZY109A ZY109B +

Lot No.: 2397
Built: 1953
G.L.W.: 19.0t   19.5t +
Tops Code: ZYV   ZYW +

LDB494009 (W) LDB494057 (W) LDB494264 (W) LDB494334+(W) LDB494596 (W)

Number Series: B496620 - B497169

Description: 13T High Goods Wagon
Builder: BR (Shildon Works)
Diagram No.: 1/146
Tare Weight: 7.0t
Design Code: ZX161A

Lot No.: 2469
Built: 1954
G.L.W.: 20.0t
Tops Code: ZXV

ADB496817

Number Series: B497620 - B498019

Description: 13T High Goods Wagon
Builder: BR (Ashford Works)   Lot No.: 2484
Diagram No.: 1/044   Built: 1954
Tare Weight: 6.5t   G.L.W.: 19.5t
Design Code: ZY109B   Tops Code: ZYW

LDB497648

Number Series: B502000 - B507549

Description: 13T Container Flat Wagon   Lot No.: 3153
Builder: Pressed Steel Co Ltd   Built: 1958
Diagram No.: 1/069   G.L.W.: 6.5t :- 9.0t +
Tare Weight: 6.0t *=#   6.5t 9.0t +   10.8t ! 20.0t $
             10.0t %   10.8t !   17.0t *=# 17.3t
Design Code: ZS009A +   ZS088D :   ZS088E -   Tops Code: ZSV =+ ZVV
             ZV083A *   ZV098A =   ZV142A   ZSO :- ZXA #
             ZX017B #   ZY015B %   ZY043B !   ZYW ^!$
             ZY302B ^

KDB503512=(W) KDB504354+(W) DB505810#   CDB506211:(W) ADB506729*(W)
ADB503515-   CDB504707=   LDB505858%   ADB506322:   CDB507266 (W)
LDB503663%(W) CDB505213 (W) KDB506009=(W) LDB506663!(W) LDB507433^
KDB504280=   KDB505757=

Number Series: B510000 - B510049

Description: 35T Bogie Container Flat Wagon
Builder: BR (Ashford Works)   Lot No.: 3494
Diagram No.: 1/075   Built: 1964
Tare Weight: 19.2t   19.75t +   G.L.W.: 36.0t 55.0t +
Design Code: FW002A + YS041A   Tops Code: FWV + YSV

   B510028+   B510037+   KDB510043   KDB510046   KDB510047
KDB510031   KDB510038

Number Series: B530000 - B530576

Description: 14T Container Flat Wagon
Builder: BR (Ashford Works)   Lot No.: 3108
Diagram No.: 1/068   Built: 1958
Tare Weight: 8.0t + 10.0t   G.L.W.: 8.0t +   10.0t
Design Code: ZS087B + ZS527C   Tops Code: ZSO   ZSW +

TDB530004+(W) ADB530199   ADB530237   ADB530331   ADB530478
ADB530051

Number Series: B530577 - B530738

Description: 14T Container Flat Wagon
Builder: BR (Derby Works)   Lot No.: 3384
Diagram No.: 1/068   Built: 1961
Tare Weight: 6.0t * 8.0t 8.6t +   G.L.W.: 8.0t 8.6t +
Design Code: ZS087B ZS527B + ZV192B *   20.0t *
Tops Code: ZSV + ZSW   ZVV *

ADB530597+   TDB530606 (W) KDB530627*(W)

Number Series: B531000 - B531017

Description: 38T Bogie Container Flat Wagon
Builder: BR (Townhill C&W)   Lot No.: 3814
Diagram No.: 1/204   Built: 1973
Tare Weight: 15.5t   G.L.W.: 54.0t
Design Code: YV050A   Tops Code: YVP

ADB531000   ADB531001   ADB531002 (W)

Number Series: B550000 - B550499

Description: 16T Mineral End Door Wagon
Builder: Birmingham R C & W Co Ltd   Lot No.: 2907
Diagram No.: 1/108   Built: 1958
Tare Weight: 7.8t   G.L.W.: 24.5t
Design Code: ZH017E   Tops Code: ZHV

 DB550116 (W)   DB550321 (W)   DB550359 (W) DB550420 (W)   DB550493 (W)
 DB550200 (W)   DB550334 (W)

Number Series: B551600 - B552949

Description: 16T Mineral End Door Wagon
Builder: Pressed Steel Co Ltd   Lot No.: 3145
Diagram No.: 1/117   Built: 1957
Tare Weight: 7.5t + 7.8t * 8.0t #   G.L.W.: 24.0t +   24.5t
             8.05t   8.5t =   25.0t =
Design Code: MC003D + MC003B # ZH017D   Tops Code: MCV +   ZHV
             ZH017E * ZY108D =   ZYV =

 DB551602*(W)   B552179#(W)   DB552419 (W)   DB552601 (W)   DB552729 (W)
 DB551604 (W)   DB552226 (W)   DB552440 (W)   DB552635 (W)   B552782+(W)
 DB551622 (W)   DB552244 (W)   DB552453 (W)   DB552637 (W)   DB552824 (W)
 B551694#(W)   DB552249 (W)   DB552510 (W) LDB552641=(W)   DB552843 (W)
 DB551766 (W)   DB552264 (W)   DB552561 (W)   DB552667 (W)   DB552897 (W)
 B551934#(W)   DB552270 (W)   DB552585 (W)   DB552711 (W)   DB552930 (W)
 DB552175 (W)   DB552295 (W)

Number Series: B554430 - B554899

Description: 16T Mineral End Door Wagon
Builder: Cravens R C & W Co Ltd
Diagram No.: 1/108
Tare Weight: 7.8t
Design Code: ZH017E

Lot No.: 2912
Built: 1958
G.L.W.: 24.5t
Tops Code: ZHV

DB554584 (W)   DB554826 (W)   DB554854 (W)

Number Series: B555250 - B555749

Description: 16T Mineral End Door Wagon
Builder: Teeside Bridge Engineering Ltd
Diagram No.: 1/108
Tare Weight: 7.8t
Design Code: ZH017E

Lot No.: 3144
Built: 1958
G.L.W.: 24.5t
Tops Code: ZHV

DB555261 (W)   DB555416 (W)   DB555493 (W)   DB555619 (W)   DB555731 (W)
DB555356 (W)   DB555428 (W)   DB555507 (W)   DB555723 (W)   DB555734 (W)
DB555409 (W)

Number Series: B556050 - B557049

Description: 16T Mineral End Door Wagon
Builder: Metropolitan Cammell C & W Co Ltd
Diagram No.: 1/108
Tare Weight: 7.8t 9.0t +
Design Code: MC001D + ZH017E

Lot No.: 2915
Built: 1957
G.L.W.: 24.5t   25.5t +
Tops Code: MCV + ZHV

DB556132 (W)   DB556325 (W)   DB556483 (W)   DB556787 (W)   DB556825 (W)
DB556225 (W)   DB556334 (W)   DB556602 (W)   DB556807 (W)   DB556907 (W)
DB556234 (W)   DB556434 (W)   B556759+(W)   DB556821 (W)   DB557026 (W)

Number Series: B557750 - B558749

Description: 16T Mineral End Door Wagon
Builder: Pressed Steel Co Ltd
Diagram No.: 1/117
Tare Weight: 8.0t 8.5t +
Design Code: MC003B * ZH017D   ZY108D +

Lot No.: 3146
Built: 1958
G.L.W.: 24.5t   25.0t +
Tops Code: MCV *   ZHV
             ZYV +

DB557757 (W)   DB558019 (W)   DB558340 (W) LDB558437+(W)   DB558485 (W)
DB557765 (W)   DB558062 (W)   DB558377 (W)   DB558468 (W)   DB558655 (W)
DB557835 (W)   B558090*(W)   DB558411 (W)   DB558480 (W)   DB558698 (W)
DB557889 (W)

Number Series: B560200 - B561199

Description: 16T Mineral End Door Wagon
Builder: Metropolitan Cammell C & W Co Ltd   Lot No.: 2917
Diagram No.: 1/108   Built: 1956
Tare Weight: 8.5t   G.L.W.: 25.0t
Design Code: ZH017A   Tops Code: ZHV

  DB561069 (W)

Number Series: B561200 - B562199

Description: 16T Mineral End Door Wagon
Builder: R Y Pickering & Co Ltd   Lot No.: 2918
Diagram No.: 1/108   Built: 1956
Tare Weight: 6.5t +   8.5t   G.L.W.: 23.0t +   25.0t
Design Code: ZH017A   ZY108A +   Tops Code: ZHV   ZYV +

LDB561292+(W)   DB561555 (W)   DB561687 (W)

Number Series: B562200 - B562799

Description: 16T Mineral End Door Wagon
Builder: Teeside Bridge Engineering Co Ltd   Lot No.: 2919
Diagram No.: 1/108   Built: 1957
Tare Weight: 8.5t   G.L.W.: 25.0t
Design Code: ZH017A   Tops Code: ZHV

  DB562324 (W)   DB562542 (W)

Number Series: B562800 - B569299

Description: 16T Mineral End Door Wagon   Lot No.: 2920
Builder: Pressed Steel Co Ltd   Built: 1956
Diagram No.: 1/108   G.L.W.: 23.0t +   25.0t
Tare Weight: 6.5t +   8.5t   9.0t *=   25.5t *=
Design Code: MC001B *   MX001A =   ZH017A   Tops Code: MCV *   MXV =
  ZY108A +   ZHV   ZYV +

DB563066 (W)	DB564344 (W)	DB565723 (W)	DB567756 (W)	LDB568742+(W)
DB563109 (W)	DB564612 (W)	B565994=(W)	DB567917 (W)	DB568901 (W)
LDB563195+	DB565018 (W)	LDB566062+(W)	DB568300 (W)	DB568974 (W)
DB563477 (W)	DB565154 (W)	DB567082 (W)	DB568335 (W)	B569153=(W)
B563591*(W)	LDB565405+(W)	DB567375 (W)	DB568377 (W)	DB569195 (W)
DB563832 (W)	DB565654 (W)	LDB567466+(W)	DB568668 (W)	DB569233 (W)

Number Series: B569300 - B576299

Description: 16T Mineral End Door Wagon
Builder: Pressed Steel Co Ltd   Lot No.: 2921
Diagram No.: 1/108   Built: 1956
Tare Weight: 8.5t   9.0t *   G.L.W.: 25.0t   25.5t *
Design Code: MX001A * ZH017A   Tops Code: MXV * ZHV

DB569487 (W)	B569624*(W)	DB570618 (W)	DB571631 (W)	DB575410 (W)
DB569513 (W)	DB569692 (W)	DB570739 (W)	B574112 (W)	DB575554 (W)
B569575*(W)	DB570474 (W)	B570855*(W)	DB574576 (W)	DB576213 (W)

Number Series: B576300 - B583299

Description: 16T Mineral End Door Wagon
Builder: Pressed Steel Co Ltd   Lot No.: 2922
Diagram No.: 1/108   Built: 1957
Tare Weight: 6.5t + 8.5t   9.0t *=   G.L.W.: 23.0t + 25.0t
Design Code: MC001C = MX001A * ZH017A   25.5t =*
             ZY108A +   Tops Code: MCV = MXV *
                                                           ZHV   ZYV +

DB576803 (W)	DB578530 (W)	TDB579284	B580835=(W)	DB581948 (W)
ADB577804 (W)	LDB578913+	B579357=(W)	DB581430 (W)	LDB581971+(W)
B578209*(W)	DB579128 (W)	DB580371 (W)	DB581644 (W)	DB582270*(W)
DB578488 (W)	DB579194 (W)	DB580399 (W)	B581841=(W)	DB582348 (W)

Number Series: B583300 - B590299

Description: 16T Mineral End Door Wagon
Builder: Pressed Steel Co Ltd   Lot No.: 2923
Diagram No.: 1/108   Built: 1957
Tare Weight: 8.0t   9.0t +*   G.L.W.: 24.5t   25.5t +*
Design Code: MC001D * ZH017E   ZY117A +   Tops Code: MCV * ZHV
                                                           ZYV +

DB587304 (W)	DB587922 (W)	DB589062 (W)	DB589743 (W)	DB589921 (W)
DB587364 (W)	LDB588438+	DB589218 (W)	DB589744 (W)	DB589947 (W)
DB587533 (W)	DB588464 (W)	DB589230 (W)	DB589827 (W)	DB589960 (W)
DB587565 (W)	DB588555 (W)	DB589266 (W)	DB589834 (W)	DB590109 (W)
DB587843 (W)	DB588564 (W)	B589329*(W)	DB589920 (W)	DB590186 (W)
DB587847 (W)	DB588799 (W)	B589496*(W)		

Number Series: B595150 - B595499

Description: 16T Mineral End Door Wagon
Builder: Derbyshire C & W Co Ltd   Lot No.: 3063
Diagram No.: 1/108   Built: 1958
Tare Weight: 8.0t   9.0t +   G.L.W.: 24.5t   25.5t +
Design Code: MC001D + ZH017E   Tops Code: MCV + ZHV

DB595155 (W)	DB595212 (W)	B595378+(W)	DB595410 (W)	DB595454 (W)
DB595160 (W)	DB595308 (W)	DB595397		

Number Series: B596000 - B596393

Description: 16T Mineral End Door Wagon
Builder: BR (Horwich Works)
Diagram No.: 1/194
Tare Weight: 8.0t +   8.8t
Design Code: MX002A +   ZH023A   ZH023B #

Lot No: 3863
Built: 1975-76 & 1978
G.L.W.: 24.5t
Tops Code: MXV +   ZHV

DB596025#(W)   DB596151 (W)   B596183+(W)   DB596263 (W)   DB596373 (W)
DB596084 (W)   B596173+(W)   DB596234 (W)   DB596333 (W)   DB596380 (W)
DB596150 (W)

Number Series: B700000 - B700499

Description: 11T Container Flat Wagon
Builder: BR (Swindon Works)
Diagram No.: 1/067
Tare Weight: 6.0t
Design Code: ZX172A

Lot No.: 2688
Built: 1955
G.L.W.: 17.5t
Tops Code: ZXV

KDB700309

Number Series: B700500 - B700999

Description: 11T Container Flat Wagon
Builder: BR (Derby Works)
Diagram No.: 1/067
Tare Weight: 6.0t
Design Code: ZV164B

Lot No.: 2836
Built: 1956
G.L.W.: 17.5t
Tops Code: ZVV

KDB700630

Number Series: B701000 - B701449

Description: 11T Container Flat Wagon
Builder: BR (Derby Works)
Diagram No.: 1/067
Tare Weight: 6.0t
Design Code: ZR118A

Lot No.: 2837
Built: 1956
G.L.W.: 17.0t
Tops Code: ZRV

ADB701441 (W)

Number Series: B702500 - B703499

Description: 11T Container Flat Wagon
Builder: BR (Ashford Works)
Diagram No.: 1/067
Tare Weight: 8.0t
Design Code: ZS086B

Lot No.: 2854
Built: 1956
G.L.W.: 8.0t
Tops Code: ZSX

TDB702702 (W)

Number Series: B704500 - B705499

Description: 13T Container Flat Wagon
Builder: BR (Swindon Works)
Diagram No.: 1/069
Tare Weight: 6.0t
Design Codes: ZV083A   ZV098A +

Lot No.: 2971
Built: 1957
G.L.W.: 17.0t
Tops Code: ZVV

ADB705090   KDB705454+

Number Series: B705500 - B706199

Description: 13T Container Flat Wagon
Builder: BR (Swindon Works)
Diagram No.: 1/069
Tare Weight: 6.5t
Design Code: ZS088D

Lot No.: 2972
Built: 1957
G.L.W.: 6.5t
Tops Code: ZSO

ADB705871

Number Series: B706200 - B706699

Description: 13T Container Flat Wagon
Builder: BR (Derby Works)
Diagram No.: 1/069
Tare Weight: 6.0t
Design Code: ZV083A

Lot No.: 2984
Built: 1957
G.L.W.: 17.0t
Tops Code: ZVV

ADB706470 (W)

Number Series: B706700 - B707199

Description: 13T Container Flat Wagon
Builder: BR (Derby Works)
Diagram No.: 1/069
Tare Weight: 6.0t
Design Code: ZV098A

Lot No.: 2985
Built: 1957
G.L.W.: 17.0t
Tops Code: ZVV

KDB707030

Number Series: B707200 - B708199

Description: 13T Container Flat Wagon
Builder: BR (Ashford Works)
Diagram No.: 1/069
Tare Weight: 6.0t
Design Code: ZR119A

Lot No.: 3024
Built: 1958
G.L.W.: 17.0t
Tops Code: ZRV

KDB707917 (W)

Number Series: B708200 - B709199

Description: 13T Container Flat Wagon
Builder: BR (Ashford Works)
Diagram No.: 1/069
Tare Weight: 6.0t
Design Code: ZR119A + ZV098A

Lot No.: 3025
Built: 1957
G.L.W.: 17.0t
Tops Code: ZRV + ZVV

KDB708849   KDB708457+(W) KDB708933

Number Series: B709200 - B709549

Description: 13T Container Flat Wagon
Builder: BR (Derby Works)
Diagram No.: 1/069
Tare Weight: 6.5t
Design Code: ZS088E

Lot No.: 3083
Built: 1958
G.L.W.: 6.5t
Tops Code: ZSQ

ADB709220

Number Series: B709700 - B709949

Description: 13T Container Flat Wagon
Builder: BR (Wolverton Works)
Diagram No.: 1/069
Tare Weight: 6.0t
Design Code: ZV142A

Lot No.: 3084
Built: 1958
G.L.W.: 17.5t
Tops Code: ZVV

KDB709799

Number Series: B709950 - B710249

Description: 13T Container Flat Wagon
Builder: BR (Swindon Works)
Diagram No.: 1/069
Tare Weight: 6.5t
Design Code: ZS088D

Lot No.: 3096
Built: 1959
G.L.W.: 6.5t
Tops Code: ZSO

CDB710066

Number Series: B710250 - B710255

Description: 22T Pallet Van
Builder: BR (Ashford Works)
Diagram No.: 1/235
Tare Weight: 10.0t
Design Code: ZS310A

Lot No.: 3362
Built: 1961
G.L.W.: 10.0t
Tops Code: ZSV

DB710253 (W)

MSV, B388896 is pictured at Thornton Junction on 1st August 1989.
Paul W. Bartlett

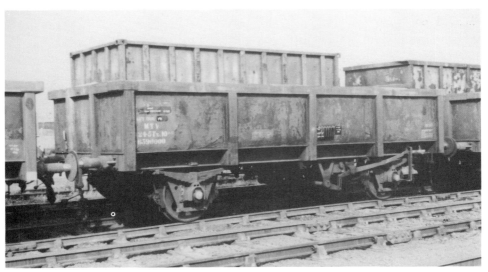

The entire class of 24T Open Sand Wagons have been transfered to departmental stock and the large majority remain in use. B390000 is pictured before transfer at Middlesborough on 23rd September 1988.
Paul W. Bartlett

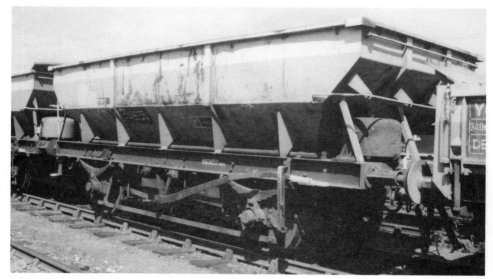

Prior to its conversion and transfer to departmental service in 1986 DB423760, ZDV, was a Coal Hopper Wagon. It is seen at Perth on 1st August 1989.
Paul W. Bartlett

ADB531000 was converted from a Rectank in 1973. More recently it has been further converted for departmental use and is pictured in its current guise on 3rd August 1989 at Millerhill.
Paul W. Bartlett

On 1st August 1989 ZGA DB715031, Fishkind "Seal", is pictured at Perth.
Paul W. Bartlett

All surviving members of the 20T Shock Absorbing Wagon have been transferred to departmental use. DB726326, ZCV Fishkind "Dace" was photographed at Woking on 28th June 1992.
Paul W. Bartlett

In the late 1960s and 1970s a number of 10T Car Transporter Wagons were converted from Coach Underframes. Subsequently transfered to departmental use KDB745234 is an example. Now withdrawn it is pictured at Bescot on 4th September 1993. Peter Ifold

Demountable Tank Wagon ADB749070, since withdrawn, was photographed at Perth on 1st August 1989. Paul W. Bartlett

Number Series: B715000 - B715019

Description: 21T High Goods Wagon
Builder: BR (Lancing Works)   Lot No.: 2579
Diagram No.: 1/055   Built: 1957
Tare Weight: 11.0t #=   11.5t   11.55t *   G.L.W.: 14.0t +   32.5t #=
             14.0t +   33.0t
Design Code: ZG084C *   ZG084D #   ZG084F =   Tops Code: ZGA *#=   ZVX :
             ZV224A :   ZX162A   ZX162B -   ZXB +   ZXX
             ZX162C +

KDB715002+	ADB715007	ADB715010	ADB715014	ADB715018
KDB715005*	KDB715008=	ADB715011	ADB715015:	KDB715019*
ADB715006-	DB715009#	KDB715012=	DB715016#(W)	

Number Series: B715020 - B715039

Description: 21T High Goods Wagon   Seal.
Builder: BR (Lancing Works)   Lot No.: 2851
Diagram No.: 1/055   Built: 1957
Tare Weight: 11.0t =#   11.5t +*   11.55t   G.L.W.: 32.5t #=   33.0t
Design Code: ZG084B *   ZG084C   ZG084D #   Tops Code: ZGW *   ZXX +
             ZG084E :   ZG084F =   ZX162A +   ZGA

KDB715020	DB715025#	KDB715030*	KDB715034=	DB715037:
ADB715021	ADB715026+	DB715031#	KDB715035	ADB715038
KDB715022=	ADB715027	KDB715032=	KDB715036=	KDB715039
KDB715023	KDB715028			

Number Series: B725675 - B725874

Description: 12T High Goods Wagon
Builder: BR (Derby Works)   Lot No.: 3232
Diagram No.: 1/056   Built: 1959
Tare Weight: 8.2t   G.L.W.: 20.0t
Design Code: ZG025B   Tops Code: ZGV

KDB725694(W)

Number Series: B726225 - B726524

Description: 20T Shock Absorbing Wagon
Builder: BR (Derby Works)   Lot No.: 3429
Diagram No.: 1/058   Built: 1962
Tare Weight: 10.5t   Fishkind: "DACE"   G.L.W.: 31.0t
Design Code: SU001A +   ZC512A   Tops Code: SUW +   ZCV

DB726225	DB726239	DB726255	DB726269 (W)	DB726289
DB726228	DB726240 (W)	DB726257	DB726270 (W)	DB726292
DB726230 (W)	DB726242	DB726258 (W)	DB726271 (W)	DB726293 (W)
DB726231 (W)	DB726244	DB726259	DB726275	DB726297 (W)
DB726232	DB726245	DB726261 (W)	DB726277	DB726298
DB726234 (W)	DB726246 (W)	DB726263	DB726281	DB726299
B726235+(W)	DB726251	DB726266	DB726288 (W)	DB726300 (W)

DB726302	DB726346	DB726389	DB726440	DB726481
DB726303	DB726347	DB726390 (W)	DB726441	DB726482
.B726306+(W)	DB726353	DB726395	DB726443	DB726484
DB726307	DB726355	DB726396	DB726446	DB726485
DB726308	DB726357	DB726398	DB726448	DB726487
DB726309	DB726358	DB726399	DB726449	DB726490
DB726312	DB726359 (W)	DB726400 (W)	DB726451	DB726492 (W)
DB726314	DB726361 (W)	DB726402	DB726453	DB726493
DB726317	DB726362	DB726403 (W)	DB726454 (W)	DB726495
DB726319	DB726363	DB726406	DB726456 (W)	DB726496
DB726320	DB726364	DB726407	DB726459	DB726497
DB726323	DB726365	DB726409 (W)	DB726461	DB726498
DB726325	DB726366	B726413+(W)	DB726462	DB726501
DB726326	DB726368	DB726417	B726463+(W)	DB726502
DB726330	DB726369	DB726418	DB726465	DB726504 (W)
DB726331	DB726370 (W)	DB726419	DB726466	B726505+(W)
DB726333 (W)	DB726371	DB726425	DB726467 (W)	DB726507
DB726334	DB726374	DB726426	DB726470	DB726511
DB726336	DB726375	DB726427	DB726471	DB726512
DB726339	DB726378	DB726428	DB726472	DB726515
DB726340	DB726379 (W)	DB726429 (W)	DB726475	DB726516
DB726341 (W)	DB726380	DB726430	DB726476	DB726517
DB726342	DB726382 (W)	DB726431	DB726477	DB726519
DB726343	DB726383 (W)	DB726436	DB726479	DB726522
DB726344	DB726384	DB726438	DB726480	DB726524

Number Series: B730000 - B730099

Description: 13.0T Low Goods Wagon
Builder: Faverdale C & W Co Ltd
Diagram No.: 1/445
Tare Weight: 10.5t #*+= 10.55t
Design Code: ZA011A   ZD107A #   ZD107B *
            ZD107C +   ZD107D =

Lot No.: 2048
Built: 1949
G.L.W.: 31.0t
Tops Code: ZAV ZDV +*#
           ZDW =

KDB730001#	KDB730016=(W)	KDB730030=(W)	KDB730051#(W)	KDB730090=
DB730003	KDB730017#(W)	KDB730035#	KDB730061*(W)	KDB730091=
KDB730005#(W)	KDB730019=	KDB730039#(W)	KDB730068=	KDB730092#(W)
KDB730006#(W)	KDB730026+(W)	KDB730046=	KDB730070=(W)	KDB730093#
KDB730008#(W)	KDB730027#(W)	KDB730048=(W)	KDB730073*(W)	KDB730094#(W)

Number Series: B730100 - B730499

Description: 20T Tube Wagon (With Full Length Doors)
Builder: Faverdale C & W Co Ltd
Diagram No.: 1/445
Tare Weight: 10.5t   10.55t :
Design Code: ZA011A -   ZA011B :   ZD107A
            ZD107B *   ZD107C !   ZD107D =
            ZV053B +

Lot No.: 2049
Built: 1950
G.L.W.: 31.0t
Tops Code: ZAV :-   ZDV
           ZVV +

KDB730100 (W)	KDB730132	KDB730151 (W)	KDB730182#(W)	KDB730194#
KDB730110	KDB730134*	KDB730155 (W)	DB730186:	KDB730198#
KDB730112*	KDB730135	KDB730158#	KDB730189	KDB730204#
KDB730129 (W)	KDB730140 (W)	KDB730161*	KDB730192#	KDB730205
KDB730131 (W)	KDB730142	KDB730164	KDB730193 (W)	KDB730212

```
KDB730215#     KDB730272      KDB730316 (W) KDB730383      KDB730430 (W)
KDB730219 (W) KDB730273!(W) KDB730317#(W) KDB730384#(W) KDB730433 (W)
KDB730222 (W) KDB730275#     KDB730318 (W) KDB730386#(W) KDB730435*
 DB730225 (W) KDB730280 (W) KDB730325*(W) KDB730388#(W) KDB730440#(W)
KDB730227      KDB730281 (W) KDB730328#(W) KDB730391      KDB730442#(W)
KDB730228#      DB730282#(W) KDB730329*     KDB730400#     KDB730443#
KDB730231 (W) KDB730286      KDB730330 (W) KDB730401*(W) KDB730446 (W)
KDB730234+(W) KDB730291#     KDB730338 (W) KDB730405 (W) KDB730448 (W)
KDB730241 (W) KDB730293      KDB730339      KDB730408 (W) KDB730450#
KDB730242*(W) KDB730296%(W) KDB730351 (W) KDB730411 (W) KDB730453#
KDB730243#      KDB730302 (W) KDB730352 (W) KDB730413 (W) KDB730467#
KDB730250 (W) KDB730304      KDB730358#     KDB730420 (W) KDB730468 (W)
 DB730255 (W) KDB730305      KDB730360*(W) KDB730421 (W) KDB730470#
KDB730256#(W) KDB730311      KDB730368 (W) KDB730425 (W) KDB730471 (W)
KDB730259      KDB730312 (W) KDB730374 (W) KDB730426 (W) KDB730472
 DB730263 (W) KDB730313 (W)  DB730377 (W) KDB730427 (W) KDB730479 (W)
KDB730268 (W) KDB730315
```

Number Series: B730500 - B730919

Description: 22T Tube Wagon
Builder: BR (Derby Works)   Lot No.: 3288
Diagram No.: 1/448   Built: 1960
Tare Weight: 11.4t =:   11.45t   11.5t *a   G.L.W.: 34.0t
Design Code: ST009J a   ST009L *   ZA009B < Tops Code: STV "a ZGV !+
             ZA010A -   ZA010B >   ZD105C =   ZAV -<>
             ZD105E :   ZD113A #   ZD113B   ZDV   ZRV %
             ZG086A !   ZG086B +   ZR188A %

```
 DB730515<(W)  DB730607-(W) ADB730738#    KDB730800+(W) KDB730863=
ADB730530#    ADB730617 (W)  B730753a(W)  DB730811<(W) ADB730866#(W)
KDB730564 (W) ADB730630#(W) ADB730756#(W) KDB730812%(W) ADB730868#(W)
KDB730568 (W) ADB730633 (W) ADB730773!(W) KDB730826      KDB730873+(W)
ADB730575+     B730652a(W) KDB730781      ADB730831#(W) KDB730889>
KDB730579+    ADB730665#(W)  DB730784-(W) KDB730837+     KDB730897+(W)
 DB730581<(W) ADB730681#(W)  B730785*     KDB730848+(W) ADB730901 (W)
ADB730593:    ADB730693#(W) ADB730788#(W) KDB730851+(W)  DB730902<(W)
 DB730597<(W) KDB730704 (W) KDB730791=     DB730852-(W) ADB730907#(W)
ADB730600=    ADB730732#(W)
```

Number Series: B730920 - B730999

Description: 22T Tube Wagon
Builder: BR (Swindon Works)   Lot No.: 3332
Diagram No.: 1/448   Built: 1961
Tare Weight: 11.45t   G.L.W.: 34.0t
Design Code: ZA009F -   ZA010A *   ZD113A   Tops Code: ZAV *-   ZDV
             ZG086B +   ZS136C #   ZGV +   ZSW #

```
KDB730926+(W) KDB730953*(W) ADB730958 (W)  DB730986      KDB730995 (W)
ADB730944 (W)  DB730954-(W)  DB730977*(W) ADB730987 (W)
```

Number Series: B731100 - B731389

Description: 22T Steel Tube Wagon
Builder: Faverdale C & W Co Ltd   Lot No.: 2204
Diagram No.: 1/447   Built: 1951
Tare Weight: 16.0t   G.L.W.: 16.0t
Design Code: ZS116A   Tops Code: ZSO

ADB731348 (W)

Number Series: B731590 - B732039

Description: 22T Steel Tube Wagon   Lot No.: 2554
Builder: Faverdale C & W Co Ltd   Built: 1954
Diagram No.: 1/448   G.L.W.: 30.0t "   34.0t
Tare Weight: 7.5t " 9.5t a   11.45t   11.5t <#
Design Code: ST009J <   ZA009E +   ZA010A =   Tops Code: STV <   ZAV +=
             ZD105A "   ZD105B a   ZD105C   ZDA #   ZYV !
             ZD113B -   ZD142A #   ZG085A >   ZGV >*%^   ZDV
             ZG086B ^   ZG087B %   ZY101B !   Fishkind: "COD" =

ADB731597#   ADB731704"(W) KDB731796=(W) KDB731867*(W) DB731957=(W)
 DB731600+(W) KDB731707^   DB731810=(W) ADB731868-(W) DB731980+(W)
KDB731609*(W) ADB731711-(W) ADB731818"   KDB731869*(W) DB731984=(W)
 DB731615=(W) LDB731727!(W) KDB731820*(W) DB731876=(W) ADB731997%(W)
 DB731617=(W) KDB731741*   KDB731821>(W) KDB731879*   DB731998-(W)
ADB731622-   KDB731751-(W) DB731823+(W) DB731889=(W) DB732009%(W)
ADB731624*(W) ADB731753-(W) KDB731828+(W) KDB731905*(W) DB732012a(W)
ADB731627-   ADB731756"   DB731832=(W) ADB731906:(W) DB732013=(W)
ADB731640+(W) ADB731758-(W) KDB731843%(W) KDB731914*(W) ADB732017-
 B731644<(W)   DB731763=(W) KDB731846*   B731915<(W) ADB732018-
KDB731645*(W) KDB731768%(W) KDB731848^(W) DB731925=(W)   B732021=(W)
ADB731655-(W) ADB731774"(W) DB731850=   DB731927=(W) ADB732023-
ADB731669-   KDB731779=   ADB731852-(W) DB731933=(W) ADB732024=
 DB731690 (W) KDB731784=(W) DB731853=(W) ADB731936%(W) KDB732027"
 DB731694=(W) DB731785=(W) KDB731859-(W) ADB731937a(W) KDB732039#
ADB731695-   DB731786=(W)

Number Series: B732040 - B732389

Description: 22T Tube Wagon   Lot No.: 2740
Builder: Faverdale C & W Co Ltd   Built: 1955
Diagram No.: 1/448   Fishkind: "COD" =   G.L.W.: 11.5t ^   12.0t <
Tare Weight: 8.5t b 11.4t !+% 11.45t   30.0t !   31.0t b
             11.5t ^za 12.0t<   34.0t
Design Code: ST009B z   ST009H b   ZA009A #   Tops Code: STV bz ZSW ^<
             ZA009E "   ZA010A *   ZD105A !   ZAV #"* ZYV a
             ZD105B +   ZD113A >   ZD113B   ZDV   ZGV :=%
             ZG086A x   ZG086B =   ZG087B :   ZSW ^<   ZVV x
             ZG087D %   ZS136C ^   ZS147A < ZY101D a

ADB732063   KDB732093x   DB732110*(W) KDB732125=(W) ADB732147
ADB732065   ADB732102   DB732117*(W) ADB732137 (W) ADB732149+(W)
ADB732067+(W) KDB732103=   DB732118*(W) B732138z(W) KDB732150:(W)
KDB732088+(W) ADB732105 (W)   DB732119*   KDB732141=(W) ADB732151!
KDB732089#   KDB732109%   DB732120"(W) ADB732146   ADB732156+(W)

                              42

```
ADB732157=(W)  ADB732224>(W)  ADB732264+(W)  KDB732300%(W)  KDB732349:(W)
ADB732160!     ADB732227+(W)  ADB732266+(W)  KDB732317  (W)  KDB732352+(W)
ADB732180!      DB732229%     ADB732283+(W)   DB732324*(W)  KDB732360!(W)
 DB732187*     ADB732234      ADB732285+(W)  KDB732325=(W)   DB732362*
ADB732189+(W)  ADB732238%     ADB732286+(W)  ADB732327+     KDB732366=
ADB732191+(W)  KDB732244:     LDB732287a(W)  ADB732328+(W)   DB732367"(W)
ADB732205+(W)   DB732251^      DB732288"(W)  ADB732330      KDB732370!
ADB732206*      DB732252<     ADB732291+     KDB732332:(W)  KDB732373%
KDB732209=(W)  KDB732253=     ADB732296      ADB732338  (W)  ADB732385+(W)
 DB732215"(W)  KDB732254+(W)   B732297b(W)   ADB732341*(W)
```

Number Series: B732390 - B733039

```
Description: 22T Tube Wagon
Builder: Faverdale C & W Co Ltd              Lot No.: 2867
Diagram No.: 1/448        Fishkind: "COD" *  Built: 1956
Tare Weight: 8.5t y   11.4t !+%              G.L.W.: 30.0t !   31.0t y
             11.45t   11.5t xz                       34.0t
Design Code: ST009C z   ST009H y   ST009J x  Tops Code: STV xyz
             ZA009E ^   ZA010A *   ZD105A !             ZAV ^*   ZVV "<
             ZD105B +   ZD113A >   ZD113B                ZDV     ZGV :%=
             ZG086B =   ZG087B :   ZG087D %
             ZV217A "   ZV218B <
```

```
ADB732398+(W)  KDB732529%(W)  KDB732694%(W)  ADB732798!      DB732930*(W)
ADB732403+(W)  KDB732558"     ADB732699  (W)  ADB732821+(W)  KDB732936=
 DB732414^(W)  ADB732595>(W)  ADB732707+(W)  ADB732825!     ADB732946+
ADB732415!     ADB732599+(W)  ADB732710+(W)  ADB732833  (W)  KDB732951=(W)
KDB732416%(W)  ADB732602+(W)   B732711y(W)   ADB732844^(W)  KDB732958^(W)
 B732419x(W)   KDB732605:(W)   DB732713^(W)  ADB732856+(W)  ADB732964
KDB732430!     ADB732617>(W)  KDB732719:(W)  KDB732857:      DB732975^(W)
ADB732432  (W) KDB732622=(W)  ADB732721  (W)  B732864z(W)   KDB732976%
KDB732433+(W)  ADB732625+(W)  ADB732726+(W)  KDB732866!     KDB732982%
KDB732436:     KDB732627>(W)  KDB732727%(W)  ADB732867!(W)  ADB732990
 DB732447*(W)  ADB732629+(W)  ADB732728>(W)  ADB732873>(W)   DB733000^
ADB732449>(W)   B732635x(W)    DB732734^(W)  ADB732883+(W)  KDB733009:(W)
KDB732452%(W)  ADB732638>(W)  ADB732741+(W)  ADB732884!(W)  ADB733012
ADB732479+(W)  ADB732641!(W)  ADB732744!     KDB732887%     KDB733014=(W)
KDB732485%(W)  KDB732658%(W)  ADB732750+(W)  ADB732888+(W)  ADB733021+
ADB732486>(W)   DB732660*(W)  ADB732751>(W)  ADB732900+(W)  KDB733026:(W)
ADB732487+(W)  ADB732672+(W)   DB732756^(W)  KDB732903%(W)  ADB733027
ADB732498+(W)  KDB732674+(W)  ADB732761>(W)  ADB732914+     ADB733028+(W)
ADB732508  (W) KDB732676=(W)  KDB732778%     KDB732915>(W)  KDB733029%(W)
ADB732513  (W) KDB732681=(W)  ADB732783!      DB732923^(W)  KDB733031>(W)
KDB732519+(W)   DB732685^(W)  KDB732793<(W)  KDB732925+     KDB733038=(W)
KDB732525+     ADB732690+(W)  KDB732797=(W)  ADB732928+(W)  ADB733039  (W)
KDB732526:(W)
```

Number Series: B733040 - B733219

Description: 22T Tube Wagon   Lot No.: 3226
Builder: Faverdale C & W Co Ltd   Built: 1959
Diagram No.: 1/448   Fishkind: "COD" *   G.L.W.: 11.5t = 34.0t
Tare Weight: 11.4t 11.45t :*+ 11.5t -=   Tops Code: ZAV :*   ZDB -
Design Code: ZA009B :   ZA010A *   ZD105B #   ZDV   ZSW =
             ZD105E   ZD113A +   ZD113B !   ZGV >"%
             ZD142A -   ZG086A >   ZG087A "
             ZG087D %   ZS136C =

KDB733040+(W)   DB733066:.(W) ADB733104 (W) ADB733136 (W)   DB733171:(W)
KDB733042#(W) ADB733068   ADB733108!(W)   DB733139:(W) KDB733176"
ADB733045-   ADB733070!(W) ADB733111 (W) KDB733145%   ADB733177!(W)
 DB733046*(W) DB733074:(W) ADB733114 (W) ADB733151 (W) KDB733197"(W)
KDB733054 (W) ADB733081 (W) ADB733116 (W) ADB733156 (W) ADB733205
 DB733057=   DB733087:(W) DB733121:(W) KDB733166:(W) ADB733206 (W)
ADB733062 (W) ADB733093 (W) KDB733127"   ADB733168 (W) KDB733208"(W)
ADB733063   KDB733098"(W) KDB733129>(W) KDB733169 (W) KDB733214>

Number Series: B733220 - B733239

Description: 22T Tube Wagon
Builder: Faverdale C & W Co Ltd   Lot No.: 3258
Diagram No.: 1/449   Built: 1961
Tare Weight: 13.6t 14.0t +   G.L.W.: 40.0t
Design Codes: ZD132A ZS149A +   Tops Code: ZSX +   ZDX

ADB733220 (W)   DB733223+(W) ADB733228   ADB733234   ADB733238
ADB733221   ADB733225   ADB733230   ADB733235   ADB733239
ADB733222   KDB733227 (W) ADB733232

Number Series: B733240 - B733459

Description: 22T Tube Wagon
Builder: BR (Derby Works)   Lot No.: 3332
Diagram No.: 1/448   Fishkind: "COD" *   Built: 1961
Tare Weight: 11.4t =+: 11.45t   11.5t !   G.L.W.: 11.5t = 34.0t
Design Code: ZA009D '   ZA009F *   ZA010A #   Tops Code: ZAV '*   ZDV
             ZD105D :   ZD105F =   ZD113A   ZGV ->+ ZYV !
             ZD113B "   ZG086B -   ZG087C >
             ZG087E +   ZY101C !

ADB733249:(W) ADB733308"(W) ADB733354:(W) KDB733379+(W) KDB733437+(W)
ADB733258=(W) DB733314*(W) ADB733356 (W) ADB733392:   ADB733439=(W)
KDB733261+(W) ADB733315:(W) DB733358*(W) ADB733399:(W) ADB733440:(W)
ADB733265 (W) KDB733316+(W) KDB733361>   KDB733402+   ADB733442:(W)
ADB733269=(W) ADB733323 (W) KDB733364+   KDB733404+(W) ADB733443=(W)
KDB733277-(W) DB733331*(W) KDB733367-   KDB733410:(W) ADB733448:
KDB733286=(W) DB733339=(W) LDB733370'(W) ADB733411=(W) KDB733456:(W)
 DB733295#(W) ADB733340:(W) KDB733374-(W) KDB733414+(W) ADB733459:(W)
ADB733301:(W) LDB733353!(W) ADB733376 (W) ADB733416"

44

Number Series: B733500 - B733999

Description: 14T Container Flat Wagon
Builder: BR (Ashford Works)   Lot No.: 2671
Diagram No.: 1/066   Built: 1954
Tare Weight: 7.0t   9.0t +   G.L.W.: 9.0t +   21.0t
Design Code: ZS085C + ZV204A   Tops Code: ZSV + ZVV

KDB733503 (W)	KDB733585 (W)	KDB733692 (W)	CDB733762+(W)	KDB733854 (W)
KDB733507 (W)	KDB733588	KDB733694	KDB733771 (W)	KDB733871 (W)
KDB733524 (W)	KDB733625 (W)	KDB733712 (W)	KDB733779 (W)	KDB733876 (W)
KDB733525 (W)	KDB733639	KDB733720	KDB733785 (W)	KDB733919 (W)
KDB733526 (W)	KDB733641 (W)	KDB733726 (W)	KDB733810 (W)	KDB733932 (W)
KDB733539 (W)	KDB733674 (W)	KDB733728	KDB733824 (W)	KDB733949 (W)
KDB733566 (W)	KDB733676 (W)	KDB733740 (W)	KDB733827 (W)	KDB733975 (W)
KDB733576	KDB733687 (W)	KDB733748 (W)		

Number Series: B734000 - B734299

Description: 14T Container Flat Wagon
Builder: BR (Swindon Works)   Lot No.: 2764
Diagram No.: 1/068   Built: 1956
Tare Weight: 8.0t   8.6t +   G.L.W.: 8.0t   8.6t +
Design Code: ZS087B * ZS087F ZS527B +   Tops Code: ZSV + ZSW

ADB734039   ADB734060+   TDB734073*(W)

Number Series: B734300 - B734559

Description: 14T Container Flat Wagon   Lot No.: 2973
Builder: BR (Swindon Works)   Built: 1957
Diagram No.: 1/068   G.L.W.: 6.0t # 8.0t +
Tare Weight: 6.0t   8.0t +   20.0t
Design Code: ZS087D # ZS087E + ZV192B   Tops Code: ZSO #+ ZVV

KDB734332 (W) KDB734351 (W) KDB734406 (W) CDB734425#   KDB734462 (W)
KDB734338 (W) ADB734363+

Number Series: B737600 - B737999

Description: 13T Container Flat Wagon (Wooden Body)
Builder: BR (Ashford Works)   Lot No.: 3107
Diagram No.: 1/069   Built: 1959
Tare Weight: 6.0t   G.L.W.: 17.0t
Design Code: ZV098A   Tops Code: ZVV

KDB737730   KDB737812   KDB737943

Number Series: B738500 - B738872

Description: 14T Container Flat Wagon
Builder: BR (Ashford Works)   Lot No.: 2489
Diagram No.: 1/064   Built: 1953
Tare Weight: 7.0t   9.0t +   G.L.W.: 9.0t +   21.0t
Design Code: ZS083B +   ZV205A   Tops Code: ZSV +   ZVV

KDB738507   KDB738601   KDB738658 (W) KDB738718   KDB738788 (W)
KDB738515 (W) KDB738612 (W) KDB738663 (W) KDB738722 (W) KDB738807 (W)
KDB738551 (W) KDB738625 (W) KDB738667   KDB738749   ADB738815+
KDB738580 (W) KDB738653   KDB738710 (W) KDB738783   KDB738836 (W)
KDB738587

Number Series: B738873 - B738942

Description: 13T Container Flat Wagon
Builder: BR (Swindon Works)   Lot No.: 2870
Diagram No.: 1/068   Built: 1956
Tare Weight: 6.0t   G.L.W.: 6.0t +   20.0t
Design Code: ZS087D +   ZV192B   Tops Code: ZSV +   ZVV

KDB738885 (W) CDB738886+

Number Series: B740000 - B740299

Description: 13T Pipe Wagon
Builder: BR (Derby Works)   Lot No.: 2004
Diagram No.: 1/460   Fishkind: "COD" *   Built: 1949
Tare Weight: 7.8t #   8.4t   G.L.W.: 8.4t *   20.0t #
Design Code: ZD109A   ZD109B -   ZD109E +   20.4t =   20.5t
             ZD109G % ZD109F =   ZD109K :   Tops Code: ZDV   ZDW =:-
             ZX167B * ZY118A #   ZXV *   ZYV #

 DB740004+   KDB740047 (W) KDB740120 (W) KDB740165 (W) KDB740241-
 DB740006-(W) KDB740050 (W) KDB740129   KDB740167-   KDB740242 (W)
 DB740007-(W) KDB740051 (W) KDB740134 (W) KDB740169-   DB740247+
ADB740011 (W) DB740059-(W) KDB740135-   KDB740182-(W) KDB740255
ADB740013 (W) KDB740063 (W) KDB740136 (W) KDB740191 (W) KDB740256%(W)
KDB740015   DB740065*(W) KDB740139-(W) KDB740198 (W) KDB740269-(W)
KDB740017 (W) KDB740068-   KDB740140 (W) LDB740200#(W) KDB740275
KDB740019 (W) KDB740075   KDB740146   KDB740211 (W) KDB740277 (W)
KDB740024=   KDB740080   KDB740147 (W) DB740218 (W) KDB740281
KDB740033-   KDB740083-   KDB740150+   ADB740220 (W) LDB740283#(W)
KDB740036=   KDB740084   KDB740153 (W) KDB740224 (W) DB740290+(W)
KDB740042-   KDB740092 (W) KDB740154:(W) KDB740231-(W) KDB740296-(W)
KDB740044 (W) KDB740101-   KDB740159 (W) KDB740233 (W) KDB740299 (W)

Number Series: B740300 - B740399

Description: 13T Pipe Wagon
Builder: Faverdale C & W Co Ltd   Lot No.: 2046
Diagram No.: 1/461   Built: 1949
Tare Weight: 8.7t   G.L.W.: 20.5t   20.7t +
Design Code: ZD118A   ZD118D =   ZD118E +   Tops Code: ZDV   ZDW =+
             ZD118F *

```
KDB740302+     KDB740326*(W) KDB740341=     KDB740378+     KDB740381 (W)
KDB740305 (W)  KDB740333+    KDB740360      KDB740379 (W)  KDB740386 (W)
KDB740318+     KDB740334=
```

Number Series: B740400 - B740599

```
Description: 13T Pipe Wagon
Builder: Faverdale C & W Co Ltd              Lot No.: 2047
Diagram No.: 1/461                           Built: 1949
Tare Weight: 7.0t :  8.7t                    G.L.W.: 19.0t #:  20.5t
Design Code: ZD118A    ZD118B #  ZD118D =           20.7t +=
             ZD118E +  ZD118G :              Tops Code: ZDV   ZDW =+:
```

```
KDB740405+     KDB740426 (W) KDB740466+     KDB740512      KDB740553 (W)
KDB740407      KDB740431+    KDB740481:     KDB740525+     KDB740554+
KDB740408+     KDB740437+    KDB740490      KDB740535+     KDB740570 (W)
KDB740409:     KDB740447 (W) KDB740492=     KDB740538      KDB740580 (W)
KDB740412 (W)  ADB740448     ADB740495 (W)  KDB740540      DB740591+
KDB740413=     DB740453+     KDB740498      KDB740541=     KDB740594+
KDB740418= °   KDB740454+(W) KDB740504 (W)  KDB740542+     KDB740596 (W)
KDB740419 (W)  DB740465+     DB740508  (W)  KDB740550#(W)
```

Number Series: B740600 - B740649

```
Description: 13T Pipe Wagon
Builder: Cambrian Wagon & Eng Co Ltd         Lot No.: 2305
Diagram No.: 1/460                           Built: 1951
Tare Weight: 8.4t                            G.L.W.: 20.4t *  20.5t
Design Code: ZD109A    ZD109B +  ZD109F *    Tops Code: ZDV   ZDW +*
```

```
KDB740601      KDB740613*    KDB740626      KDB740639 (W)  DB740645+(W)
KDB740606      KDB740617+    KDB740636+     KDB740644 (W)  KDB740647
KDB740610+
```

Number Series: B740650 - B740699

```
Description: 13T Pipe Wagon
Builder: BR (Swindon Works)                  Lot No.: 2329
Diagram No.: 1/460                           Built: 1953
Tare Weight: 8.4t                            G.L.W.: 20.4t *  20.5t
Design Code: ZD109A    ZD109B +  ZD109F *    Tops Code: ZDV   ZDW +*
```

```
KDB740653 (W)  DB740662+     KDB740671 (W) KDB740681+      KDB740688 (W)
KDB740655 (W)  KDB740665 (W) KDB740673 (W) KDB740684+      KDB740694+(W)
KDB740657 (W)  KDB740667+    KDB740675+    KDB740685+      KDB740698 (W)
KDB740660+(W)  KDB740668*    KDB740676*    KDB740687 (W)   KDB740699+
KDB740661 (W)  KDB740669+(W) KDB740678+
```

Number Series: B740700 - B740899

Description: 13T Pipe Wagon
Builder: BR (Wolverton Works)   Lot No.: 2458
Diagram No.: 1/460   Built: 1953
Tare Weight: 7.0t - 7.8t = 8.4t   G.L.W.: 19.0t - 20.0t =
Design Code: ZD109A   ZD109B + ZD109D -   20.4t * 20.5t
             ZD109E : ZD109F * ZD109G %   Tops Code: ZDV   ZDW +*
             ZY118A =   ZYV =

```
 DB740700:(W) KDB740734 (W) KDB740778     KDB740822+    KDB740868
KDB740705+    LDB740735=    KDB740782+     KDB740825     KDB740872+(W)
KDB740706+    KDB740740+    LDB740794=(W)  KDB740827     KDB740874+(W)
KDB740710     KDB740746 (W) KDB740801+(W)  KDB740830+(W) KDB740876*
KDB740712%(W) KDB740747 (W) KDB740803      KDB740841 (W) KDB740877+
ADB740713     KDB740749+    KDB740804+     KDB740843     KDB740880+(W)
KDB740714 (W) KDB740752+    KDB740807      KDB740845+    KDB740882 (W)
KDB740715 (W) KDB740754*    KDB740808      KDB740846+    KDB740887+
KDB740717     KDB740767+(W) KDB740815-     KDB740848+(W) KDB740893+
KDB740725 (W) KDB740770+    KDB740817+(W)  KDB740854+(W) KDB740894+(W)
KDB740726*    KDB740771 (W) KDB740819+     KDB740857 (W) KDB740897 (W)
KDB740728+    KDB740777     KDB740820      KDB740863+
```


Number Series: B740900 - B741099

Description: 13T Pipe Wagon
Builder: BR (Wolverton Works)   Lot No.: 2545
Diagram No.: 1/460   Built: 1954
Tare Weight: 7.0t #   8.4t   G.L.W.: 19.0t # 20.4t *
Design Code: ZD109A   ZD109B + ZD109F *   20.5t
             ZD109J # ZD109K "   Tops Code: ZDV   ZDW +*"

```
KDB740901+(W) KDB740936+(W) KDB740979*     KDB741041 (W) KDB741063 (W)
KDB740908*    KDB740937 (W) KDB740984 (W)  KDB741042     KDB741070"(W)
KDB740914 (W) KDB740942+    KDB740988 (W)  KDB741044 (W) KDB741072+
KDB740915 (W) KDB740946"(W)  DB740990+(W)  KDB741045 (W) KDB741073
KDB740917+    KDB740947+    KDB741000#     KDB741046+    KDB741077
KDB740918#    KDB740951#    KDB741003*     KDB741047     KDB741080+
KDB740919+    KDB740954     KDB741014 (W)  KDB741052     KDB741086 (W)
KDB740922 (W) KDB740959+    KDB741023 (W)  KDB741054 (W) KDB741094 (W)
KDB740923 (W) KDB740971     KDB741024+(W)  KDB741059"(W)  DB741097+(W)
KDB740928+    KDB740974+    KDB741038+(W)  ADB741060+    -KDB741098 (W)
KDB740934 (W) KDB740978+(W) KDB741040      ADB741061 (W)
```


Number Series: B741100 - B741449

Description: 12T Pipe Wagon
Builder: BR (Wolverton Works)   Lot No.: 2712
Diagram No.: 1/462   Built: 1955
Tare Weight: 8.4t   8.5t *   G.L.W.: 20.5t
Design Code: ZD108A   ZD108D : ZD108E =   Tops Code: ZDV   ZDW =+
             ZD108F + ZD117A # ZV221A *   ZVV *

```
KDB741100 (W) KDB741104+     KDB741121+     KDB741132 (W) KDB741146+
KDB741101     KDB741106      DB741122 (W)  KDB741136 (W) KDB741153+
KDB741103+    KDB741110 (W) KDB741127 (W)  KDB741143+    KDB741156
```

48

```
DB741161:    DB741213:    KDB741262+   KDB741329 (W) KDB741392+(W)
KDB741163 (W) KDB741214 (W) KDB741267+   KDB741336+(W) KDB741394
KDB741165=   KDB741217=   KDB741271+   KDB741337 (W) KDB741396#(W)
KDB741167+   KDB741220 (W) KDB741272=   KDB741338 (W) KDB741397 (W)
KDB741174+(W) DB741222 (W) KDB741287=(W) KDB741339+    KDB741400 (W)
KDB741183    KDB741231 (W) KDB741289    KDB741341 (W) KDB741401
KDB741185    KDB741232    KDB741294 (W) KDB741356 (W) KDB741402 (W)
KDB741188    KDB741238*(W) KDB741297=   KDB741359+    KDB741410
KDB741189 (W) KDB741240+   KDB741300    KDB741360+(W) KDB741420+
KDB741198 (W) KDB741244+(W) KDB741305=(W) KDB741365    ADB741422 (W)
KDB741199    KDB741245+   KDB741306+(W) KDB741370+    KDB741423 (W)
KDB741204 (W) KDB741250#(W) KDB741309+   KDB741378    KDB741441=
KDB741209    KDB741251+   KDB741312    KDB741381    KDB741445
KDB741210    KDB741252    KDB741321 (W) KDB741383+   KDB741447#
KDB741211+(W) KDB741254 (W) KDB741327 (W) KDB741389 (W) KDB741449=
KDB741212 (W) KDB741260+
```

Number Series: B741500 - B741559

```
Description: 12T Pipe Wagon
Builder: BR (Wolverton Works)                Lot No.: 2845
Diagram No.: 1/462                           Built: 1956
Tare Weight: 8.4t                            G.L.W.: 20.5t
Design Code: ZD108A    ZD108F +              Tops Code: ZDV    ZDW +
```

```
KDB741504    KDB741522 (W) KDB741531+   KDB741537+(W) KDB741549
KDB741512 (W) KDB741526 (W) KDB741535 (W) KDB741542 (W) KDB741553+(W)
KDB741519+
```

Number Series: B741560 - B741639

```
Description: 12T Pipe Wagon
Builder: BR (Wolverton Works)                Lot No.: 2846
Diagram No.: 1/462                           Built: 1957
Tare Weight: 8.4t    8.5t #                  G.L.W.: 20.5t
Design Code: ZD108A    ZD108E =   ZD108F +   Tops Code: ZDV    ZDW =+*
             ZD117A :   ZD159B *   ZV221A #              ZVV #
```

```
KDB741560#(W) KDB741586    DB741602+   KDB741618    KDB741635+
KDB741567 (W) KDB741587 (W) KDB741604+(W) KDB741620    KDB741636+
 DB741570+   KDB741589=   KDB741607 (W) KDB741623:   KDB741637:(W)
 DB741574+   KDB741591+   KDB741608*(W) KDB741627    KDB741638+
KDB741581    KDB741592:
```

Number Series: B741640 - B741669

```
Description: 12T Pipe Wagon (High Goods)
Builder: BR (Wolverton Works)                Lot No.: 2846
Diagram No.: 1/462                           Built: 1957
Tare Weight: 8.4t                            G.L.W.: 20.5t
Design Code: ZD108A    ZD108F +   ZD108K *   Tops Code: ZDV    ZDW +*
```

```
 DB741641+(W) KDB741646+(W) KDB741653 (W) KDB741660    KDB741663*(W)
KDB741642 (W) KDB741647 (W)  DB741655+(W) KDB741662+    KDB741664 (W)
```

Number Series: B741670 - B741729

Description: 12T Pipe Wagon
Builder: BR (Wolverton Works)
Diagram No.: 1/462
Tare Weight: 7.0t # 8.4t
Design Code: ZD108A   ZD108C # ZD108E =
             ZD108F + ZY118C :

Lot No.: 2846
Built: 1957
G.L.W.: 19.0t #: 20.5t
Tops Code: ZDV   ZDW =+
           ZYV :

```
  B741673  (W)  KDB741682+(W)  KDB741687  (W)  LDB741704:(W)  KDB741716  (W)
KDB741675       KDB741683+     KDB741692=      KDB741708#(W)  KDB741717  (W)
KDB741677  (W)  KDB741686=     KDB741694       KDB741715=     KDB741723  (W)
```

Number Series: B741730 - B741749

Description: 12T Pipe Wagon
Builder: BR (Wolverton Works)
Diagram No.: 1/463
Tare Weight: 8.4t   9.0t *
Design Code: ZD126A   ZD126D * ZD126E +

Lot No.: 3167
Built: 1958
G.L.W.: 20.5t 21.0t *
Tops Code: ZDV   ZDW +*

```
KDB741733*    KDB741735 (W) KDB741745      KDB741747+    KDB741748*
```

Number Series: B741750 - B741949

Description: 12T Pipe Wagon
Builder: BR (Wolverton Works)
Diagram No.: 1/463
Tare Weight: 8.4t :#*+=! 8.5t >% 8.7t < 10.9t
Design Code: ZD108F :   ZD108H > ZD108J %
             ZD117A   ZD117B # ZD117C *
             ZD117D +   ZD117E = ZD117F ! ZY118E <

Lot No.: 3335
Built: 1961
G.L.W.: 20.5t   20.7t <
Tops Code: ZDV   ZDW :%+=
           ZYV <

```
KDB741776=     DB741808:(W)  KDB741836  (W)  KDB741863=      KDB741903%(W)
KDB741778=     KDB741812>(W) KDB741837  (W)  KDB741867  (W)  KDB741904>(W)
KDB741779 (W)  KDB741815=(W) DB741842=(W)    DB741876=(W)    KDB741908
KDB741780+     KDB741819  (W) LDB741848<     KDB741884  (W)  KDB741919=(W)
KDB741782 (W)  KDB741820  (W) KDB741851=     KDB741893=(W)   KDB741926  (W)
KDB741794 (W)  KDB741822=(W) KDB741858=      KDB741895       KDB741932!(W)
KDB741799 (W)  KDB741824     KDB741860=(W)   KDB741897  (W)  LDB741945<(W)
KDB741801=(W)  KDB741831  (W) KDB741862%(W)  DB741901*       DB741948=(W)
KDB741805 (W)  KDB741833#(W)
```

Number Series: B745000 - B745053

Description: 10T Car Transporter Flat Wagon (ex Mk1 Coach Underframe)
Builder: BR (Cowlairs)
Diagram No.: 1/189
Tare Weight: 23.0t   29.3t !
G.L.W.: 29.0t :- 33.0t   39.3t !
Design Code: FV003X + FV004A = FV004F
             YR020A " YR020E * YR030A <
             YY043G : YY043J - YY043K !

Lot No.: 3679
Built: 1968-9
Tops Code: FVA   FVX +
           YRV "* YRX <
           YYA !   YYX :-

```
B745000        KDB745008*(W)  KDB745018*(W)  LDB745033-      DB745050"(W)
B745001+       LDB745009:     KDB745027<      B745035        KDB745051*(W)
B745002        LDB745011:     KDB745029*(W)   B745036        B745052=(W)
B745003        LDB745014-     KDB745031*(W)   LDB745040!     B745053
B745006+       KDB745017*(W)  KDB745032*(W)   KDB745044*(W)  KDB745056*(W)
```

Number Series: B745057 - B745061

Description: 10T Car Transporter Flat Wagon (ex Mk1 Coach Underframe)
Builder: BR (Cowlairs)   Lot No.: 3715
Diagram No.: 1/189   Built: 1969-70
Tare Weight: 23.0t *   23.5t   G.L.W.: 33.0t *   33.5t
Design Code: FV004E   FV004F +   Tops Code: FVA

```
B745057 (W)    B745058        B745059        B745060        B745061*
```

Number Series: B745062 - B745087

Description: 10T Car Transporter Flat Wagon (ex Mk1 Coach Underframe)
Builder: BR (Cowlairs)   Lot No.: 3758
Diagram No.: 1/131   Built: 1970-1
Tare Weight: 21.0t +*   23.0t =   G.L.W.: 29.0t =   31.0t +*
             23.2t   23.5t #:   33.2t   33.5t #:
Design Code: FV003V # FV004C - FV004D   Tops Code: FVA   FVX #
             FV004E : YR021E * YV071B +   YRV *   YVV +
             YY043H =   YYX =

```
KDB745062+(W)   B745073*(W)   B745077#       B745081        B745085%
KDB745063*(W)   B745074-      B745078:       B745082:       B745086:(W)
LDB745067=      B745075:      B745080        B745083:       B745087%
KDB745072*(W)   B745076:
```

Number Series: B745088 - B745227

Description: 10T Car Transporter Flat Wagon (ex Mk1 Coach Underframe)
Builder: BREL (Swindon Works)   Lot No.: 3831
Diagram No.: 1/177   Built: 1972-4
Tare Weight: 23.0t <!-   23.5t   G.L.W.: 33.0t <!-   33.5t
Design Code: YR020B <   YR023B > YR023C % Tops Code: FVW +   ZRV <
             YR023D *   FV003Z + YV073A   YXW -   YYV :
             YV074A !   YX054A - YY043A :   YVV   YRW >%*

```
 B745088+(W)    B745113+(W)   KDB745135*(W)  LDB745150:(W)  KDB745167*
KDB745090 (W)   B745115+(W)   DB745136%(W)    B745152+(W)    B745168+(W)
DB745091-       B745116+(W)    B745137+(W)    B745153+(W)   DB745170*(W)
 B745093+(W)   KDB745119*(W)  KDB745138*(W)   B745154+(W)   KDB745171*
KDB745098*      B745120+(W)   DB745139%(W)    B745156+(W)   KDB745172*
DB745100%      KDB745121*(W)   B745142+(W)   DB745159%       B745173+(W)
KDB745105*     KDB745122 (W)  DB745144%      DB745160%(W)    B745175+(W)
KDB745107*(W)   B745125+(W)    B745145+(W)   KDB745161+      B745177+(W)
 B745109*       B745128+(W)   KDB745146*      B745162+(W)   KDB745178*(W)
KDB745110*      B745129+(W)    B745147+(W)   KDB745163*      B745179+(W)
KDB745111*      B745130+(W)   KDB745148*     DB745164>       B745180+(W)
KDB745112*(W)   B745131+(W)    B745149+(W)    B745165+(W)    B745181+(W)
```

51

```
KDB745182*      B745192+(W)     B745200+(W)     B745210+(W)     B745219+(W)
KDB745183 (W)   B745193+(W)     B745203+(W)   KDB745211*        B745222+(W)
  B745184+(W)  KDB745194+(W)   KDB745204*(W)    B745213+(W)     B745223+(W)
 DB745186<(W)    B745196+(W)   KDB745205!(W)   KDB745215*(W)    B745225+(W)
  B745188+(W)    B745197+(W)    DB745206%      B745216+(W)    KDB745226*(W)
  B745189+(W)    B745199+(W)   KDB745207*       B745217+(W)
```

Number Series: B745228 - B745297

Description: 10T Car Transporter Flat Wagon (ex Mk1 Coach Underframe)
Builder: BREL (Swindon Works)   Lot No.: 3867
Diagram No: 1/177   Built: 1975
Tare Weight: 23.5t   G.L.W.: 33.5t
Design Code: FV003Z + YR023C - YR023D   Tops Code: YRW   FVW +

```
KDB745228+(W)  KDB745241 (W)  KDB745256 (W)    B745271+(W)     B745285+(W)
 DB745229-       B745242+(W)  KDB745257       KDB745273        KDB745286 (W)
  B745230+(W)  KDB745243 (W)  KDB745258        DB745274-        DB745287-
KDB745231 (W)    B745244+(W)    B745259+(W)  KDB745276          B745288+(W)
KDB745232        B745245+(W)  KDB745260        B745277+(W)    KDB745289
KDB745233 (W)  KDB745247        B745261+(W)    B745279+(W)     DB745290-
KDB745234 (W)    B745248+(W)  KDB745263        B745280+(W)     B745291+(W)
KDB745236      KDB745249 (W)    B745264+(W)    B745281+(W)    KDB745292
KDB745237 (W)   DB745250-       B745265+(W)  KDB745282 (W)     B745295+(W)
KDB745238 (W)   DB745251-(W)    B745266+(W)    DB745283-      KDB745296 (W)
 B7452239+(W)  KDB745253        B745267+(W)  KDB745284        KDB745297
  B745240+(W)    B745254+(W)  KDB745269 (W)
```

Number Series: B745298 - B745302

Description: 10T Car Transporter Flat Wagon (ex Mk1 Coach Underframe)
Builder: BREL (Doncaster Works)   Lot No.: 3868
Diagram No.: 1/177   Built: 1973
Tare Weight: 23.5t   G.L.W.: 33.5t
Design Code: FV003Z *   YRO23D   Tops Code: FVW *   YRW

```
   B745299*(W)    B745300*(W) KDB745301
```

Number Series: B745600 - B745647

Description: 10T Car Transporter Flat Wagon (ex GWR Coach Underframe)
Builder: BR (Swindon Works)   Lot No.: 3536
Diagram No.: 1/131   Built: 1964-5
Tare Weight: 21.0t   G.L.W.: 21.0t #   31.0t
Design Code: FV004B >   YR021A   YR021D *   Tops Code: FVA >   YRV
             YR021F -   YS036A # YV071A %   YRW *   YVV +%
             YV071D +   YY043C !   YSV #   YVV !

 DB745601 (W)   DB745617*   DB745623   KDB745633%(W) KDB745638-(W)
KDB745606-(W) LDB745618!   KDB745624-(W) KDB745635-(W)   DB745642+
ADB745615+(W) ADB745619#   KDB745630-(W)   B745636>   ADB745647<(W)
```
```

Number Series: B745790 - B745803

Description: 10T Car Transporter Flat Wagon (ex LMS Coach Underframe)
Builder: BR (St Rollox Works)   Lot No.: 3563
Diagram No.: 1/129   Built: 1965
Tare Weight: 23.5t   G.L.W.: 33.5t
Design Code: FV011B   Tops Code: FVX

  B745800

Number Series: B745846 - B745865

Description: 10T Car Transporter Flat Wagon (ex LMS Coach Underframe)
Builder: BR (Derby Works)   Lot No.: 3588
Diagram No.: 1/132   Built: 1966
Tare Weight: 22.5t *+   23.5t   G.L.W.: 32.5t *+   33.5t
Design Code: FV010B * FV010F YR029A +
Tops Code: FVA FVX * YRA +

  B745846   B745850   B745858 (W) KDB745862+(W)   B745863
  B745847*   B745852   B745859 (W)

Number Series: B745866 - B745871

Description: 10T Car Transporter Flat Wagon (ex LMS Coach Underframe)
Builder: BR (St Rollox Works)
        BR (Derby Works) +   Lot No.: 3589
Diagram No.: 1/132   Built: 1966
Tare Weight: 22.5t   G.L.W.: 33.5t
Design Code: FV011B   Tops Code: FVX

  B745866   B745867 (W)   B745868+(W)   B745871

Number Series: B745967 - B745986

Description: 10T Bogie Electrification Flat Wagon (ex GWR Coach
            Underframe)
Builder: BR (Swindon Works)   Lot No.: 3641
Diagram No.: 1/088   Built: 1967
Tare Weight: 23.5t   G.L.W.: 33.5t
Design Code: YY044A   Tops Code: YYV

LDB745975 (W)

Number Series: B748110 - B748129

Description: 20T Open Carriage Wagon
Builder: BR (Ashford Works)   Lot No.: 2770
Diagram No.: 1/092   Fishkind: "BREAM" Built: 1958
Tare Weight: 13.0t +   13.1t *   15.0t   G.L.W.: 15.0t 33.0t +*
Design Code: ZE005A   ZX148B + ZX148C * Tops Code: ZEX ZXX +*

```
DB748110 (W)    DB748115       DB748119       DB748124       DB748127
DB748111        DB748116       DB748120+      DB748125       DB748128
DB748112 (W)    DB748117       DB748121       DB748126*      DB748129
DB748114        DB748118       DB748122
```

Number Series: B748130 - B748149

```
Description: 20T Open Carriage Wagon
Builder: BR (Ashford Works)                    Lot No.: 2850
Diagram No.: 1/092        Fishkind: "BREAM"    Built: 1958
Tare Weight: 13.1t +   15.0t                   G.L.W.: 15.0t    33.0t +
Design Code: ZE005A    ZX148C +                Tops Code: ZEX   ZXX +
```

```
DB748130        DB748134       DB748138       DB748142          DB748147+
DB748131        DB748135       DB748139       DB748143          DB748148
DB748132        DB748136       DB748141       DB748144 (W)      DB748149
DB748133        DB748137
```

Number Series: B748698 - B748722

```
Description: 10T Car Transporter Flat Wagon (ex LNER Coach Underframe)
Builder: BR (Ashford Works)                    Lot No.: 3533
Diagram No.: 1/088                             Built: 1964
Tare Weight: 23.5t                             G.L.W.: 33.5t
Design Code: YR024B    YRO24C %   YR028A =     Tops Code: YRA =    YRV %
             YV070A -                                     YRW      YVV -
```

```
DB748701      KDB748702-(W) KDB748703%(W)  DB748708       KDB748720=
```

Number Series: B748723 - B748747

```
Description: 10T Car Transporter Flat Wagon (ex LNER Coach Underframe)
Builder: BR (Horwich Works)                    Lot No.: 3534
Diagram No.: 1/129                             Built: 1964
Tare Weight: 23.5t                             G.L.W.: 33.5t
Design Code: FV011C                            Tops Code: FVA
```

```
B748728         B748730 (W)
```

Number Series: B748748 - B748768

```
Description: 10T Car Transporter Flat Wagon (ex LNER Coach Underframe)
Builder: BR (Derby Works)                      Lot No.: 3535
Diagram No.: 1/129                             Built: 1964
Tare Weight: 23.5t                             G.L.W.: 33.5t
Design Code: FV011C                            Tops Code: FVA
```

```
B748755
```

Number Series: B749023 - B749032

Description: Demountable Tank Wagon
Builder: BR (Derby Works)
Diagram No.: 1/327
Tare Weight: 7.0t
Design Code: ZR121A

Lot No.: 2073
Built: 1950
G.L.W.: 13.0t
Tops Code: ZRV

ADB749028 (W) ADB749030 (W)

Number Series: B749066 - B749070

Description: Demountable Tank Wagon
Builder: Earlestown C & W Co Ltd
Diagram No.: 1/334
Tare Weight: 7.0t
Design Code: ZR123A

Lot No.: 2946
Built: 1956
G.L.W.: 13.0t
Tops Code: ZRV

ADB749070 (W)

Number Series: B749200 - B749203

Description: Demountable Tank Wagon
Builder: BR (Shildon Works)
Diagram No.: 1/325
Tare Weight: 7.0t
Design Code: ZR529A

Lot No.: 2044
Built: 1950
G.L.W.: 13.0t
Tops Code: ZRV

ADB749203

Number Series: B749400 - B749407

Description: Demountable Tank Wagon
Builder: BR (Derby Works)
Diagram No.: 1/329
Tare Weight: 8.0t
Design Code: ZS093A

Lot No.: 2077
Built: 1950
G.L.W.: 8.0t
Tops Code: ZSV

TDB749406 (W)

Number Series: B749412 - B749417

Description: Demountable Tank Wagon
Builder: Earlestown C & W Co Ltd
Diagram No.: 1/337
Tare Weight: 7.0t
Design Code: ZR124A

Lot No.: 2427
Built: 1953
G.L.W.: 13.0t
Tops Code: ZRV

DB749416 (W)

Number Series: B749600 - B749609

Description: Tank Wagon
Builder: Grazebrook C & W Co Ltd   Lot No.: 2170
Diagram No.: 1/301   Built: 1951
Tare Weight: 13.5t   G.L.W.: 31.0t
Design Code: ZR107A   Tops Code: ZRP

  DB749601 (W)   DB749604   DB749608   DB749609

Number Series: B749650 - B749659

Description: 20T Tank Wagon
Builder: BR (Ashford Works)   Lot No.: 2240
Diagram No.: 1/304   Built: 1951
Tare Weight: 10.5t   11.8t * 12.0t +   G.L.W.: 31.0t   32.3t *
Design Code: ZR101B   ZR101C + ZR101D *   32.5t +
             ZR108B #   Tops Code: ZRR   ZRP *

ADB749650#(W) ADB749652 (W)   DB749653+(W) DB749658*

Number Series: B749660 - B749679

Description: 20T Tank Wagon
Builder: BR (Ashford Works)   Lot No.: 2429
Diagram No.: 1/305   Built: 1953
Tare Weight: 12.0t   G.L.W.: 32.5t
Design Code: ZR102A + ZR120B   Tops Code: ZRR   ZRQ +

  DB749671+(W)   DB749673+(W)   DB749676 (W) DB749678   DB749679
  DB749672+(W)   DB749675 (W)

Number Series: B750000 - B751299

Description: 12T Ventilated Goods Van (LMS Design)
Builder: BR (Wolverton Works)   Lot No.: 2001
Diagram No.: 1/200   Built: 1949
Tare Weight: 8.0t   G.L.W.: 20.0t
Design Code: ZD008B   ZR025A +   Tops Code: ZDV   ZRV +

KDB750268 (W)   DB750504+(W)

Number Series: B751300 - B751799

Description: 12T Ventilated Goods Van (LMS Design)
Builder: BR (Wolverton Works)   Lot No.: 2003
Diagram No.: 1/204   Built: 1949
Tare Weight: 7.9t + 8.0t   G.L.W.: 19.0t + 20.0t
Design Code: ZD010A   ZP502B + ZR027A *
Tops Code: ZDV   ZPV + ZRV *

KDB751313 (W)   DB751405*(W)   DB751786+(W)

Number Series: B751800 - B752349

Description: 12T Ventilated Goods Van (LMS Design)
Builder: BR (Wolverton Works)   Lot No.: 2013
Diagram No.: 1/204   Built: 1949
Tare Weight: 7.9t   G.L.W.: 7.9t
Design Code: ZP502B   Tops Code: ZPW

DB752335

Number Series: B752350 - B752789

Description: 12T Ventilated Goods Van
Builder: BR (Ashford Works)   Lot No.: 2062
Diagram No.: 1/202   Built: 1949
Tare Weight: 7.5t   G.L.W.: 19.0t
Design Code: ZR026A   Tops Code: ZRV

KDB752730 (W)

Number Series: B752790 - B753099

Description: 12T Ventilated Goods Van
Builder: BR (Ashford Works)   Lot No.: 2063
Diagram No.: 1/202   Built: 1949
Tare Weight: 7.5t   G.L.W.: 19.0t
Design Code: ZQ002A   Tops Code: ZQV

KDB752868 (W)

Number Series: B753430 - B754429

Description: 12T Ventilated Goods Van
Builder: BR (Wolverton Works)   Lot No.: 2109
Diagram No.: 1/204   Built: 1950
Tare Weight: 7.9t   8.0t *   G.L.W.: 19.0t   20.0t *
Design Code: ZP502A   ZR027A *   Tops Code: ZPV   ZRV *

  DB753637 (W) KDB753722*(W) ADB753792*

Number Series: B755180 - B756679

Description: 12T Ventilated Goods Van
Builder: BR (Wolverton Works)   Lot No.: 2181
Diagram No.: 1/208   Built: 1951
Tare Weight: 7.25t   G.L.W.: 19.0t
Design Code: ZD012A   Tops Code: ZDV

KDB756128 (W)

Number Series: B756680 - B758179

Description: 12T Ventilated Goods Van
Builder: BR (Wolverton Works)
Diagram No.: 1/208
Tare Weight: 6.0t   7.6t *
Design Code: ZQ007A * ZY095A

Lot No.: 2182
Built: 1951
G.L.W.: 19.0t
Tops Code: ZQV *   ZYV

DB757283*(W) LDB757301

Number Series: B758180 - B759179

Description: 12T Ventilated Goods Van
Builder: BR (Wolverton Works)
Diagram No.: 1/208
Tare Weight: 7.25t
Design Code: ZD012A

Lot No.: 2318
Built: 1952
G.L.W.: 19.0t
Tops Code: ZDV

ADB758372 (W)

Number Series: B760880 - B761343

Description: 12T Ventilated Goods Van
Builder: Faverdale C & W Co Ltd
Diagram No.: 1/208
Tare Weight: 7.25t +   7.6t
Design Code: ZD012A +   ZQ007A

Lot No.: 2421
Built: 1954
G.L.W.: 19.0t
Tops Code: ZDV +   ZQV

KDB761241+(W)   DB761319 (W)

Number Series: B761380 - B762179

Description: 12T Ventilated Goods Van
Builder: BR (Wolverton Works)
Diagram No.: 1/208
Tare Weight: 7.6t
Design Code: ZR029A

Lot No.: 2465
Built: 1954
G.L.W.: 19.0t
Tops Code: ZRV

KDB761677 (W)   DB761319 (W)

Number Series: B762280 - B762429

Description: 12T Ventilated Goods Van
Builder: Faverdale C & W Co Ltd
Diagram No.: 1/212
Tare Weight: 7.4t
Design Code: ZQ010A

Lot No.: 2585
Built: 1954
G.L.W.: 19.0t
Tops Code: ZQV

DB762324 (W)

Number Series: B762450 - B763279

Description: 12T Ventilated Goods Van
Builder: BR (Wolverton Works)
Diagram No.: 1/208
Tare Weight: 7.0t   7.25t *
Design Code: ZD012A *   ZQ007A +   ZR029A

Lot No.: 2595
Built: 1954
G.L.W.: 19.0t
Tops Code: ZDV *   ZQV +
          ZRV

ADB762490 (W) KDB762579*(W) KDB762624*(W) KDB763046 (W) KDB763162 (W)
 DB762577+(W)   DB762586+   KDB762855 (W)   DB763058*   KDB763226 (W)

Number Series: B763281 - B764030

Description: 12T Ventilated Goods Van
Builder: BR (Wolverton Works)
Diagram No.: 1/208
Tare Weight: 7.0t
Design Code: ZQ007A

Lot No.: 2414
Built: 1955
G.L.W.: 19.0t
Tops Code: ZQV

 DB763773 (W)

Number Series: B764481 - B765480

Description: 12T Ventilated Goods Van
Builder: Faverdale C & W Co Ltd
Diagram No.: 1/208
Tare Weight: 7.0t   7.25t *
Design Code: ZD012A *   ZP503A =   ZQ007A +
             ZR029A

Lot No.: 2367
Built: 1952
G.L.W.: 19.0t
Tops Code: ZDV *   ZPV =
          ZQV +   ZRV

ADB764528 (W)   DB764620=(W) KDB764661*(W)   DB765011+(W)

Number Series: B765481 - B766400

Description: 12T Ventilated Goods Van
Builder: Faverdale C & W Co Ltd
Diagram No.: 1/213
Tare Weight: 7.0t   7.5t =
Design Code: ZD015A *   ZR033A +   ZV193A =
             ZY100C

Lot No.: 2422
Built: 1952
G.L.W.: 19.0t   19.5t =
Tops Code: ZDV *   ZRV +
          ZVV =   ZYV

KDB765494+(W)   DB765936=(W)   DB766153=(W)   DB766210*(W) ADB766236 (W)
ADB765671*(W) KDB766019*

Number Series: B766401 - B767600

Description: 12T Ventilated Goods Van
Builder: BR (Wolverton Works)   Lot No.: 2706
Diagram No.: 1/208   Built: 1955
Tare Weight: 7.0t   7.25t * 8.0t =   G.L.W.: 19.0t   20.0t =
Design Code: ZD012A *   ZR029A   ZR029C +   Tops Code: ZDV *   ZRV
             ZV181A =   ZRW + ZVV =

KDB766748*(W) KDB766896*(W)   DB767048=(W) KDB767219+   KDB767228(W)

Number Series: B767601 - B769400

Description: 12T Ventilated Goods Van
Builder: BR (Wolverton Works)   Lot No.: 2707
Diagram No.: 1/208   Built: 1956
Tare Weight: 7.0t   8.0t =   G.L.W.: 19.0t   20.0t =
Design Code: ZD012A *   ZQ007A +   ZR029A #   Tops Code: ZDV *   ZQV +
             ZV181A =   ZRV # ZVV =

 DB767568+(W) KDB767777   KDB767796#   DB768846=   DB768924*(W)
 DB767690*(W)   DB767786+

Number Series: B769401 - B769900

Description: 12T Ventilated Goods Van
Builder: Faverdale C & W Co Ltd   Lot No.: 2735
Diagram No.: 1/213   Built: 1957
Tare Weight: 7.0t   G.L.W.: 19.0t
Design Code: ZD015A *   ZR033A   Tops Code: ZDV *   ZRV

KDB769490 (W) ADB769566 (W) KDB769607*(W) KDB769749 (W)

Number Series: B770151 - B770650

Description: 12T Ventilated Goods Van
Builder: BR (Wolverton Works)   Lot No.: 2840
Diagram No.: 1/208   Built: 1956
Tare Weight: 7.0t   G.L.W.: 19.0t
Design Code: ZP503A ZQ007A +   Tops Code: ZPV   ZQV +

 DB770314 (W)   DB770386+(W)

Number Series: B771451 - B772200

Description: 12T Ventilated Goods Van   Lot No.: 2855
Builder: BR (Ashford Works)   Built: 1956
Diagram No.: 1/213   G.L.W.: 19.0t   19.5t +
Tare Weight: 7.0t   7.5t +   Tops Code: ZDV   ZQV *
Design Code: ZD015A ZQ011A * ZV193A +   ZVV +

 DB771766+(W)   DB771773 (W) KDB772143*(W)

60

<u>Number Series: B773351 - B774450</u>

Description: 12T Ventilated Goods Van
Builder: BR (Wolverton Works)   Lot No.: 2990
Diagram No.: 1/208   Built: 1957
Tare Weight: 7.0t   7.5t #   G.L.W.: 19.0t   20.0t #
Design Code: ZD012A = ZQ007A   Tops Code: ZDV = ZQV
             ZR029A * ZV051B #   ZRV * ZVV #

  DB773357*   KDB773390*(W) DB773513 (W) KDB773838*(W)   DB774443#
KDB773363=(W) DB773513 (W) KDB773401=(W)

<u>Number Series: B774451 - B775449</u>

Description: 12T Ventilated Goods Van
Builder: BR (Wolverton Works)   Lot No.: 2991
Diagram No.: 1/208   Built: 1957
Tare Weight: 7.0t   G.L.W.: 19.0t
Design Code: ZD012A * ZR029A   Tops Code: ZDV * ZRV

KDB774857 (W) KDB774997*(W)

<u>Number Series: B775451 - B776150</u>

Description: 12T Ventilated Goods Van
Builder: Faverdale C & W Co Ltd   Lot No.: 3007
Diagram No.: 1/208   Built: 1957
Tare Weight: 7.0t   G.L.W.: 19.0t
Design Code: ZD015A   Tops Code: ZDV

  DB775938 (W)

<u>Number Series: B776551 - B777350</u>

Description: 12T Ventilated Goods Van
Builder: BR (Ashford Works)   Lot No.: 3023
Diagram No.: 1/213   Built: 1957
Tare Weight: 7.0t   G.L.W.: 19.0t
Design Code: ZQ011A   Tops Code: ZQV

  DB776570 (W)

<u>Number Series: B777351 - B778250</u>

Description: 12T Ventilated Goods Van   Lot No.: 3086
Builder: BR (Wolverton Works)   Built: 1958
Diagram No.: 1/208   G.L.W.: 19.0t   20.0t =
Tare Weight: 7.0t   8.5t =   Tops Code: ZDV   ZDV *
Design Code: ZD012A   ZD012B * ZQ007A +   ZQV + ZRB =
             ZR193A =

  DB777356+(W)   DB777676   DB777764 (W) DB777796*(W) LDB778246=

Number Series: B778251 - B778750

Description: 12T Ventilated Goods Van
Builder: BR (Wolverton Works)
Diagram No.: 1/208
Tare Weight: 7.0t   8.5t =
Design Code: ZR033B   ZR192A =

Lot No.: 3228
Built: 1959
G.L.W.: 19.0t   20.0t =
Tops Code: ZRB =   ZRV

DB778331=(W) KDB778561 (W)

Number Series: B778751 - B779050

Description: 12T Ventilated Goods Van
Builder: BR (Wolverton Works)
Diagram No.: 1/208
Tare Weight: 10.0t
Design Code: ZR206B

Lot No.: 3191
Built: 1958
G.L.W.: 20.0t
Tops Code: ZRB

ADB779026 (W)

Number Series: B779551 - B779850

Description: 12T Ventilated Goods Van
Builder: Faverdale C & W Co Ltd
Diagram No.: 1/211
Tare Weight: 10.0t
Design Code: ZR206B

Lot No.: 3118
Built: 1958
G.L.W.: 20.0t
Tops Code: ZRB

ADB779834 (W)

Number Series: B779851 - B780550

Description: 12T Shock Absorbing Goods Van
Builder: BR (Ashford Works)
Diagram No.: 1/218
Tare Weight: 7.0t   8.5t +*
Design Code: ZD015B   ZD015C =   ZR033B :
             ZR192A +   ZR193A *   ZY100A !
             ZY100D #

Lot No.: 3109
Built: 1958
G.L.W.: 19.0t   20.0t +*
Tops Code: ZDW =   ZDV %
               ZRB +*   ZRV
               ZYV !#

ADB779896 (W) LDB780060!(W) KDB780331%(W) ADB780339#(W) LDB780465+(W)
LDB780046*(W)   DB780238=

Number Series: B780551 - B781750

Description: 12T Ventilated Goods Van
Builder: Chas Roberts & Co Ltd
Diagram No.: 1/208
Tare Weight: 7.0t #:+   8.0t *   8.5t
Design Code: ZR193A   ZX099C +   ZV051A *
             ZD012A #   ZD012B :

Lot No.: 3164
Built: 1958-59
G.L.W.: 19.0t #:+   20.0t
Tops Code: ZRB   ZXV +
               ZVV *   ZDV #:

62

LDB780575 (W) LDB781263 (W) LDB781479 (W)   DB781547#   LDB781595 (W)
LDB781007 (W) LDB781375 (W) ADB781539*(W)   DB781564:(W)   DB781680:(W)
 DB781091+

Number Series: B781752 - B782273

Description: 12T Ventilated Goods Van
Builder: BR (Wolverton Works)   Lot No.: 3310
Diagram No.: 1/217   Built: 1960
Tare Weight: 10.0t   G.L.W.: 20.0t
Design Code: RX001A + ZR206A   Tops Code: RXB + ZRB

  B781763+(W) ADB781847 (W) ADB781864 (W) ADB781875 (W)

Number Series: B782873 - B783872

Description: 12T Ventilated Goods Van
Builder: BR (Wolverton Works)   Lot No.: 3391
Diagram No.: 1/217   Built: 1962
Tare Weight: 8.0t   G.L.W.: 20.0t
Design Code: ZD146A # ZR214A * ZR214C % Tops Code: ZDV # ZRV *
             ZX163B + ZY107A   ZQ091A >   ZRW % ZXW +
                                                       ZYV   ZQV >

KDB782960%(W) KDB783093*(W) LDB783287   DB783392>   KDB783688%(W)
ADB783056+(W) KDB783176%(W) KDB783354%(W) DB783643#(W) DB783864*(W)
KDB783082%(W)   DB783248#(W) KDB783364#(W)

Number Series: B783873 - B784772

Description: 12T Ventilated Goods Van
Builder: BR (Derby Works)   Lot No.: 3392
Diagram No.: 1/234   Built: 1962
Tare Weight: 8.0t   G.L.W.: 20.0t
Design Code: ZR214A   ZR214C : ZY107A +   Tops Code: ZRV   ZRW :
             ZQ091A *   ZYV + ZQV *

KDB783925   DB784095*(W) TDB784175 (W) KDB784199:(W) KDB784455:(W)
LDB784017+(W) DB784130*   LDB784187+(W) KDB784281:(W) DB784694*(W)

Number Series: B784873 - B786872

Description: 12T Ventilated Goods Van
Builder: Pressed Steel & Co Ltd   Lot No.: 3398
Diagram No.: 1/224   Built: 1961
Tare Weight: 8.0t   8.5t +*   G.L.W.: 20.0t
Design Code: ZY100B   ZQ092A + ZR186A : Tops Code: ZDV # ZQB +
             ZR192A * ZR214A = ZD104A #   ZRB * ZRV :=
                                                      ZYV

LDB785252   ADB786095#(W) LDB786393*(W) ADB786510:(W) ADB786600#(W)
KDB785802=(W)   B786343 (W) KDB786448#(W) ADB786586:(W) ADB786627:(W)
TDB786001+

Number Series: GB786873 - GB787022

Description: 20T Ventilated Ferry Van
Builder: Pressed Steel & Co Ltd
Diagram No.: 1/227
Tare Weight: 12.6t + 12.9t * 14.9t =^>
             15.0t   15.1t %
Design Code: RB034A " RR044A + ZR222A
             ZR222C < ZR222D % ZR223A !
             ZS143A : ZS146A - ZQ079A =
             ZQ079C > ZV226A * ZY124A ^

Lot No.: 3413
Built: 1962
G.L.W.: 12.6t + 15.0t :"
        35.4t %=^> 35.5t
        40.0t *
Tops Code: RBX " RRX +
           ZQB > ZQX =
           ZRX   ZSX :-
           ZVX * ZYX ^

KDB786873	KDB786900 (W)	KDB786931	DB786963=	KDB786993 (W)
KDB786874 (W)	DB786901-	DB786932-	DB786965-(W)	DB786994-
KDB786875=	DB786902-	DB786933=	KDB786966	KDB786995-
KDB786876-	DB786903!	KDB786934!	KDB786967 (W)	DB786996=
KDB786877 (W)	KDB786905	DB786935=	DB786968=	LDB786997*
TDB786878%	DB786906>	KDB786937!	LDB786969^	KDB786998=
DB786879-(W)	KDB786909 (W)	KDB786938	DB786971-	KDB786999<
KDB786880!	KDB786910=	KDB786939	LDB786972^	KDB787000!
KDB786881	DB786911-(W)	KDB786940	DB786973-	KDB787001
B786883	KDB786912 (W)	KDB786942	DB786975-	KDB787003
KDB786884 (W)	LDB786913^	KDB786943	KDB786976=	DB787004-
KDB786885	DB786914-	DB786944-(W)	DB786977-(W)	KDB787005:(W)
KDB786886	KDB786916=(W)	DB786946=(W)	DB786978=	KDB787006<
DB786888-(W)	KDB786917	KDB786948 (W)	KDB786979	DB787007-
DB786889>	KDB786918	KDB786949!	DB786980-	LDB787008^
KDB786890	B786919"(W)	KDB786950	KDB786981-	KDB787009=
KDB786891	DB786920-	DB786951-	KDB786982=(W)	KDB787010
KDB786892<(W)	KDB786921 (W)	DB786952=	KDB786984	DB787011-
DB786893=	DB786922-	KDB786953	DB786986-	KDB787014
DB786894-	KDB786924	KDB786954 (W)	KDB786988=	KDB787016
DB786896:(W)	DB786927-(W)	KDB786958	DB786989>	KDB787018 (W)
KDB786897	DB786928=(W)	KDB786959	KDB786990=	KDB787021
DB786898=	DB786929 (W)	B786960"(W)	DB786991-	KDB787022
DB786899=(W)	DB786930 (W)			

Number Series: GB787098 - GB787347

Description: 20T Ventilated Ferry Van
Builder: BR (Ashford Works)
Diagram No.: 1/227
Tare Weight: 10.5t   12.9t + 13.1t =
             15.0t ><#-*"
Design Code: RR043A   ZE008A % ZC516A = ZR215A # ZR215B > ZR222B -
             ZS143B * ZS153A : ZP048A x ZQ094A < ZV226A + ZY124B "
Tops Code: RRX   ZEX % ZCX = ZRX #>- ZSX *: ZPX ^ ZQX < ZVX +
           ZYX "   Fishkind: "BREAM" % "CHUBB" =

Lot No.: 3472
Built: 1963
G.L.W.: 10.5t   15.0t *
        35.5t x
        40.0t =><#-*+"

KDB787099#	B787111 (W)	KDB787123*	LDB787133+	DB787157*(W)
LDB787100+	LDB787112+	KDB787124*	DB787140*	ADB787158#(W)
DB787101*(W)	LDB787114+	KDB787125#	LDB787143+	DB787159*
B787102 (W)	KDB787115*	KDB787126-	LDB787145+	DB787160*
LDB787104+	B787116	DB787127#	DB787146*(W)	DB787163*
B787105 (W)	DB787117=(W)	B787128 (W)	KDB787147*	DB787165 (W)
DB787106*	LDB787119+(W)	KDB787129-	DB787148-	KDB787167#
B787108 (W)	DB787120*	B787130 (W)	KDB787150:(W)	ADB787169+(W)
KDB787110*(W)	B787122 (W)	KDB787132-	B787154 (W)	KDB787170*

LDB787171+	LDB787208+	B787243 (W)	ADB787284>(W)	B787315 (W)
B787173 (W)	DB787209*(W)	DB787244= —	KDB787285#	DB787316=
B787180 (W)	KDB787210#	LDB787245"	LDB787286+	LDB787317+
DB787181-	B787212 (W)	ADB787246+	DB787287=	KDB787319x
LDB787182+	B787213 (W)	LDB787248+	LDB787288+	DB787320*
LDB787183+	B787214 (W)	ADB787249+(W)	DB787289*(W)	DB787321*(W)
DB787184=	LDB787215+	LDB787250+	KDB787290*	DB787322=(W)
LDB787185+	LDB787216"	DB787251*(W)	KDB787291#(W)	B787323 (W)
DB787186%	KDB787218#	LDB787253"	ADB787293#	B787324 (W)
B787187 (W)	DB787219=	DB787254*(W)	B787294 (W)	ADB787325+
B787188 (W)	B787220 (W)	DB787257*	DB787295*	ADB787327+
KDB787189#(W)	KDB787221*	B787259 (W)	ADB787296#	B787328 (W)
DB787190*	KDB787222#(W)	DB787260=	DB787297<	DB787329*(W)
ADB787192+	DB787224*	B787261	TDB787298*(W)	ADB787330+
B787193 (W)	B787225 (W)	LDB787263+	KDB787299*	LDB787331"
LDB787195+	DB787226*	DB787266*	DB787300*	KDB787332*
DB787196*(W)	KDB787227*	KDB787271#	DB787301=(W)	DB787333*(W)
DB787197*	LDB787228+	LDB787272+	KDB787302+	DB787334-(W)
DB787198=	LDB787229+	LDB787273+	B787305 (W)	LDB787337+
DB787199=	LDB787231+	KDB787275#(W)	B787306	DB787338*
DB787201*	KDB787232#	DB787276+	DB787307*	DB787340*
KDB787202#	DB787234*	LDB787277"	DB787309*(W)	B787343 (W)
LDB787203:	KDB787235-(W)	DB787279*	KDB787310-	B787344 (W)
KDB787204#(W)	KDB787237#(W)	B787280 (W)	B787312 (W)	DB787346*
KDB787205*	DB787241#(W)	B787282	LDB787313+	LDB787347>(W)
LDB787206+	LDB787242+	B787283 (W)	LDB787314+	

Number Series: B787395

Description: 20T Vanfit (Experimental Full Length Doors)
Builder: BR (Eastleigh Works)   Lot No.: 3567
Diagram No.: 1/239   Built: 1966
Tare Weight: 15.3t   G.L.W.: 29.5t
Design Code: ZR210A   Tops Code: ZRB

ADB787395

Number Series: B787397 - B787462

Description: 22T Ventilated Pallet Van
Builder: BR (Ashford Works)   Lot No.: 3579
Diagram No.: 1/235   Fishkind: "BREAM" +   Built: 1966
Tare Weight: 12.0t +   12.2t   G.L.W.: 12.0t
Design Code: RR039A   ZE004A +   Tops Code: RRB   ZEB +

B787397 (W)	B787408	B787419 (W)	B787430	DB787441+
B787398 (W)	B787409	DB787420+	B787431 (W)	DB787442+
B787399	B787410	DB787421+	B787432	B787443 (W)
B787400	B787411 (W)	DB787422+	B787433	B787444 (W)
B787401 (W)	DB787412+	B787423	B787434	DB787445+
B787402 (W)	DB787413+	B787424 (W)	B787435 (W)	DB787446+
B787403 (W)	B787414	DB787425+	B787436 (W)	B787447 (W)
DB787404+	B787415	B787426	B787437 (W)	DB787448+
DB787405+	DB787416+	B787427	DB787438+	DB787449+
B787406 (W)	DB787417+	B787428 (W)	B787439 (W)	B787450
DB787407+	DB787418+(W)	B787429	B787440 (W)	DB787451+

```
  DB787452+      DB787454+      B787456      B787458      B787460
   B787453 (W)    B787455 (W)  DB787457+      B787459     DB787461+
```

Number Series: B787463 - B787478

```
Description: 22T Ventilated Pallet Van
Builder: BR (Ashford Works)                Lot No.: 3586
Diagram No.: 1/235       Fishkind: "BREAM" +  Built: 1966
Tare Weight: 12.0t +  12.2t                G.L.W.: 12.0t
Design Code: RR039A    ZE004A +            Tops Code: RRB    ZEB +

  B787464        DB787467+      B787470 (W)  DB787473+      DB787476+
   B787465 (W)   DB787468+      B787471 (W)   B787474 (W)   DB787477+
  DB787466+       B787469 (W)   B787472       B787475 (W)   DB787478+
```

Number Series: B850100 - B850599

```
Description: 12T Ventilated Goods Van
Builder: BR (Ashford Works)                Lot No.: 2158
Diagram No.: 1/207                         Built: 1950
Tare Weight: 8.7t                          G.L.W.: 20.7t
Design Code: ZQ006B                        Tops Code: ZQV

  DB850261 (W)
```

Number Series: B851600 - B852349

```
Description: 12T Ventilated Goods Van
Builder: Faverdale C & W Co Ltd            Lot No.: 2471
Diagram No.: 1/209                         Built: 1953
Tare Weight: 8.5t    9.0t +                G.L.W.: 20.5t    21.0t +
Design Code: ZQ008A   ZR030A +             Tops Code: ZQV   ZRV +

  DB852094 (W)    B852895+(W)
```

Number Series: B854526 - B855000

```
Description: 12T Ventilated Goods Van      Lot No.: 3117
Builder: Faverdale C & W Co Ltd            Built: 1959
Diagram No.: 1/218                         G.L.W.: 20.0t %    21.5t
Tare Weight: 8.0t %     9.5t               Tops Code: ZDV *   ZQV #
Design Code: ZD016A *   ZQ087A #   ZR034A +           ZRV +   ZYV
             ZY099A     ZY099B %                       ZYW %

- ADB854583+     ADB854723+(W) LDB854816 (W) LDB854880 (W) KDB854958+
  LDB854599       B854758#(W) LDB854875%(W) KDB854920*(W) KDB854998+(W)
```

Number Series: B870760 - B870879

Description: 24T Covered Hopper Wagon (Covhop)
Builder: BR (Ashford Works)   Lot No.: 3393
Diagram No.: 1/210   Built: 1963
Tare Weight: 12.3t   G.L.W.: 37.0t
Design Code: CH001L   Tops Code: CHV

  B870832 (W)

Number Series: B873024 - B873193

Description: 22T Cement Hopper Wagon
Builder: Gloucester R C & W Co Ltd   Lot No.: 3323
Diagram No.: 1/272   Built: 1960
Tare Weight: 13.2t   G.L.W.: 36.0t
Design Code: CP001B   Tops Code: CPV

  B873099 (W)   B873103 (W)   B873139 (W)

Number Series: B873570 - B873719

Description: 22T Cement Hopper Wagon
Builder: Central C & W Co Ltd   Lot No.: 3409
Diagram No.: 1/272   Built: 1962
Tare Weight: 13.2t   G.L.W.: 36.0t
Design Code: CP001B   Tops Code: CPV

  B873617 (W)   B873624 (W)   B8731687 (W)   B873717 (W)

Number Series: B873794 - B873893

Description: 22T Cement Hopper Wagon
Builder: Central C & W Co Ltd   Lot No.: 3497
Diagram No.: 1/272   Built: 1964
Tare Weight: 13.2t   G.L.W.: 36.0t
Design Code: CP001B   Tops Code: CPV

  B873848 (W)   B873868 (W)

Number Series: B881610 - B882009

Description: 12T Banana Van
Builder: BR (Wolverton Works)   Lot No.: 3209
Diagram No.: 1/246   Built: 1959
Tare Weight: 9.0t   G.L.W.: 21.0t
Design Code: ZD097A   Tops Code: ZDV

ADB881741   ADB881987

Number Series: GB889000 - GB889019

Description: 14T Ferry Motor Car Van
Builder: BR (Lancing Works)
Diagram No.: 1/291
Tare Weight: 15.0t % 15.3t   15.8t -
              16.0t =
Design Code: RB039A % ZP043A = ZR211A #
             ZQ093A   ZX159A -

Lot No.: 2848
Built: 1958
G.L.W.: 15.0t %   29.5t
                  29.5t #   30.0t =-
Tops Code: ZQX   ZPX =
           ZRX #   ZXX -
           RBX %

DB889000-	DB889004	B889009%	DB889014	DB889017-
DB889001 (W)	ADB889005=	DB889010	B889015%	DB889018
ADB889002#(W)	DB889006	DB889011	ADB889016#	DB889019-
DB889003 (W)	DB889007	ADB889012#(W)-		

Number Series: GB889020 - GB889029

Description: 14T Ferry Motor Car Van
Builder: BR (Lancing Works)
Diagram No.: 1/291
Tare Weight: 15.3t   15.8t +
Design Code: ZQ093A ZR211A # ZX159A +
Tops Code: ZQX   ZRX # ZXX +

Lot No.: 3022
Built: 1958
G.L.W.: 29.5t   29.5t #
        30.0t +

DB889020+(W)	DB889022	DB889024 (W)	DB889026+(W)	DB889028# -
DB889021#	DB889023	DB889025	DB889027	DB889029#

Number Series: GB889200 - GB889203

Description: 14T Bogie Scenery Van
Builder: BR (Eastleigh Works)
Diagram No.: 1/292
Tare Weight: 30.0t
Design Code: YR025A

Lot No.: 2849
Built: 1958
G.L.W.: 44.0t
Tops Code: YRX

ADB889201 (W) ADB889203 (W)

Number Series: B900000 - B900009

Description: 20T Flatrol Wagon
Builder: BR (Derby Works)
Diagram No.: 2/512
Tare Weight: 11.0t
Design Code: ZV006C

Lot No.: 2030
Built: 1950
G.L.W.: 31.0t
Tops Code: ZVR

DB900004   DB900006

Number Series: B900010 - B900014

Description: 20T Flatrol Wagon
Builder: BR (Lancing Works)
Diagram No.: 2/512
Tare Weight: 11.0t
Design Code: ZV006C

Lot No.: 2354
Built: 1952
G.L.W.: 31.0t
Tops Code: ZVR

DB900010   DB900011   DB900012 (W)   DB900013

Number Series: B900020 - B900022

Description: 20T Flatrol Wagon
Builder: BR (Derby Works)
Diagram No.: 2/512
Tare Weight: 11.5t
Design Code: ZX056B

Lot No.: 2613
Built: 1954
G.L.W.: 32.0t
Tops Code: ZXR

DB900022

Number Series: B900023 - B900026

Description: 20T Flatrol Wagon
Builder: BR (Derby Works)
Diagram No.: 2/512
Tare Weight: 11.0t +   11.5t
Design Code: ZV006C +   ZX056B

Lot No.: 2641
Built: 1954
G.L.W.: 31.0t +   32.0t
Tops Code: ZVR +   ZXR

DB900023   DB900024+   DB900025

Number Series: B900027 - B900036

Description: 20T Flatrol Wagon
Builder: BR (Lancing Works)
Diagram No.: 2/525
Tare Weight: 15.3t +   15.5t
Design Code: ZV189B   ZV190B +

Lot No.: 2927
Built: 1958
G.L.W.: 35.5t   35.8t +
Tops Code: ZVR

DB900027   DB900030   DB900033   DB900035   DB900036
DB900028+   DB900032   DB900034

Number Series: B900037 - B900042

Description: 20T Flatrol Wagon
Builder: BR (Lancing Works)
Diagram No.: 2/512
Tare Weight: 11.0t   11.5t +
Design Code: ZV006C   ZX056B +

Lot No.: 2945
Built: 1957
G.L.W.: 31.0t   32.0t +
Tops Code: ZVR   ZXR +

DB900037+   ADB900038   DB900039+

69

Number Series: B900044 - B900046

Description: 20T Flatrol Wagon
Builder: BR (Shildon Works)
Diagram No.: 2/528
Tare Weight: 14.5t
Design Code: ZV188B   ZX116A +

Lot No.: 3199
Built: 1959
G.L.W.: 35.0t
Tops Code: ZVR   ZXP +

DB900044   DB900045   DB900046+

Number Series: B900047 - B900048

Description: 20T Flatrol Wagon
Builder: BR (Shildon Works)
Diagram No.: 2/528
Tare Weight: 14.5t
Design Code: ZV188B

Lot No.: 2878
Built: 1959
G.L.W.: 35.0t
Tops Code: ZVR

DB900047

Number Series: B900100 - B900103

Description: 20T Flatrol Wagon
Builder: Fairfields S & E Co Ltd
Diagram No.: 2/516
Tare Weight: 13.0t
Design Code: ZV007D

Lot No.: 2218
Built: 1953
G.L.W.: 34.0t
Tops Code: ZRV

DB900100   DB900101   DB900102

Number Series: B900105 - B900107

Description: 20T Flatrol Wagon
Builder: Fairfields S & E Co Ltd
Diagram No.: 2/516
Tare Weight: 13.0t
Design Code: ZV007D

Lot No.: 2619
Built: 1955
G.L.W.: 34.0t
Tops Code: ZVR

DB900105   DB900106

Number Series: B900108

Description: 20T Flatrol Wagon
Builder: Fairfields S & E Co Ltd
Diagram No.: 2/516
Tare Weight: 13.0t
Design Code: ZV007D

Lot No.: 2936
Built: 1956
G.L.W.: 34.0t
Tops Code: ZVR

DB900108

This photograph is self explanatory as ADB749652 was used whilst in departmental service for the storage of fuel. Since withdrawn it is seen at Gloucester on 26th November 1989.   Paul W. Bartlett

Once a familiar site on BR the 12T Ventilated Goods Van is becoming increasingly scarce. This Faverdale C & W built example DB761319 dating from 1954 is pictured withdrawn at Bescot on 4th September 1993.   Peter Ifold

Another example of the 12T Ventilated Goods Van is this Derby Works built example dating from 1962. LDB784187 is also now withdrawn but was still in service when seen at Millerhill on 3rd August 1989.

Paul W. Bartlett

ZCX DB787117, Fishkind "Chubb", was once a 20T Ferry Van. Pictured on the day it was condemned DB787117 is seen at Radyr on 12th June 1990.

Paul W. Bartlett

Bearing a much closer resemblance to its original Ferry Van form than DB787117 in the previous photograph is KDB787210 pictured at Millerhill on 3rd August 1989. Paul W. Bartlett

DB889029 formerly a 14T Ferry Motor Car Van was photographed at Bescot on 4th September 1993. Peter Ifold

20T Flatrol Wagon DB900006 complete with load is seen at Woking on 28th June 1992.   Paul W. Bartlett

Since withdrawn 40T Bogie Flatrol Wagon DB900405 was photographed at Rugby on 28th April 1991.   Paul W. Bartlett

Number Series: B900109 - B900128

Description: 20T Flatrol Wagon
Builder: BR (Shildon Works)
Diagram No.: 2/530
Tare Weight: 15.1t   17.0t #
Design Code: ZV209A   ZV209B *   ZV209D #
             ZX149A +

Lot No.: 3252
Built: 1959
G.L.W.: 36.6t   38.0t #
Tops Code: ZVW   ZXW +
                      ZXW +

ADB900109#	ADB900114+	ADB900118#	ADB900120#	ADB900123
DB900110*	ADB900115#	ADB900119#	DB900122*	ADB900124!
ADB900113#				

Number Series: B900300 - B900309

Description: 20T Bogie Flatrol Wagon
Builder: BR (Derby Works)
Diagram No.: 2/510
Tare Weight: 29.5t
Design Code: YV003C

Lot No.: 2006
Built: 1950
G.L.W.: 50.0t
Tops Code: YVR

B900304

Number Series: B900310 - B900313

Description: 20T Bogie Flatrol Wagon
Builder: Cambrian Wagon & Eng Co Ltd
Diagram No.: 2/515
Tare Weight: 28.3t
Design Code: YX007B

Lot No.: 2304
Built: 1952
G.L.W.: 48.8t
Tops Code: YXR

DB900312

Number Series: B900353 - B900354

Description: 25T Bogie Flatrol Wagon'
Builder: BR (Lancing Works)
Diagram No.: 1/151
Tare Weight: 32.5t
Design Code: YV006A

Lot No.: 3303
Built: 1961
G.L.W.: 58.0t
Tops Code: YVR

B900354

Number Series: B900400 - B900409

Description: 40T Bogie Flatrol Wagon
Builder: BR (Derby Works)
Diagram No.: 2/511
Tare Weight: 30.0t
Design Code: YV010C   YV010D +

Lot No.: 2005
Built: 1950
G.L.W.: 75.0t
Tops Code: YVP   YVR +

DB900405 (W) ADB900407+

Number Series: B900410 - B900414

Description: 40T Bogie Flatrol Wagon
Builder: Head Wrightson & Co Ltd
Diagram No.: 2/511
Tare Weight: 34.4t *   34.5t   37.5t +
Design Code: YV010D   YX022B +   YX022C *

Lot No.: 2364
Built: 1954
G.L.W.: 74.9t *   75.0t
                78.0t +
Tops Code: YVR   YXP *
                YXQ +

ADB900410+(W) LDB900411+(W) ADB900412   KDB900414*

Number Series: B900415 - B900419

Description: 40T Bogie Flatrol Wagon
Builder: Head Wrightson & Co Ltd
Diagram No.: 2/524
Tare Weight: 36.0t   37.5t +   38.5t *:
Design Code: YV032C   YV032F *   YX023D :
             YX023E +

Lot No.: 2451
Built: 1954
G.L.W.: 76.5t   78.0t +
                79.0t *:
Tops Code: YVQ *   YVR
                YXO :   YXR +

ADB900415   ADB900416*   DB900417+   ADB900419:

Number Series: B900425 - B900428

Description: 40T Bogie Flatrol Wagon
Builder: Head Wrightson & Co Ltd
Diagram No.: 2/524
Tare Weight: 38.5t
Design Code: YV032F

Lot No.: 3069
Built: 1958
G.L.W.: 79.0t
Tops Code: YVQ

LDB900427

Number Series: B900429 - B900430

Description: 40T Bogie Flatrol Wagon
Builder: Teesside Bridge Engineering Ltd
Diagram No.: 2/529
Tare Weight: 25.9t +   26.5t
Design Code: YX063A   YV020C +

Lot No.: 3266
Built: 1960
G.L.W.: 66.4t +   67.0t
Tops Code: YXP   YVR +

KDB900429   ADB900430+

Number Series: B900431 - B900436

Description: 40T Bogie Flatrol Wagon
Builder: Standard Railway & Wagon Co Ltd
Diagram No.: 2/531
Tare Weight: 30.4t =   30.5t   71.5t *
Design Code: YV005A   YV005C =   YV047B *
Tops Code: YVA *   YVP   YVR =

Lot No.: 3269
Built: 1958
G.L.W.: 69.9t =   71.0t
                71.5t *

DB900431   DB900435=   DB900436*

76

Number Series: B900437 - B900440

Description: 40T Bogie Flatrol Wagon
Builder: Standard Railway & Wagon Co Ltd
Diagram No.: 2/531
Tare Weight: 30.4t +   30.5t
Design Code: YV005B   YV005C +

Lot No.: 3270
Built: 1961
G.L.W.: 69.9t +   71.0t
Tops Code: YVB   YVR +

DB900437+   DB900438+   DB900440

Number Series: B900500 - B900503

Description: 50T Bogie Flatrol Wagon
Builder: P & W McLellan Ltd
Diagram No.: 2/514
Tare Weight: 35.9t
Design Code: YX006C   YX006E +

Lot No.: 2176
Built: 1951/2
G.L.W.: 86.9t
Tops Code: YXQ +   YXR

DB900500   DB900501   DB900502+

Number Series: B900507 - B900508

Description: 50T Bogie Flatrol Wagon
Builder: BR (Lancing Works)
Diagram No.: 2/514
Tare Weight: 35.9t
Design Code: YX006F

Lot No.: 2952
Built: 1959
G.L.W.: 86.9t
Tops Code: YXR

KDB900507 (W)   DB900508 (W)

Number Series: B900600 - B900601

Description: 30T Bogie Flatrol Wagon
Builder: BR (Lancing Works)
Diagram No.: 2/521
Tare Weight: 39.5t
Design Code: YV061B

Lot No.: 2571
Built: 1955
G.L.W.: 70.0t
Tops Code: YVR

ADB900600 (W)

Number Series: B900800 - B900809

Description: 20T Bogie Weltrol Wagon
Builder: BR (Derby Works)
Diagram No.: 2/730
Tare Weight: 19.5t
Design Code: YV012A   YV012C +

Lot No.: 2029
Built: 1950
G.L.W.: 40.0t
Tops Code: YVO   YVP +

ADB900804+   ADB900808 (W)

Number Series: B900810 - B900811

Description: 20T Bogie Weltrol Wagon
Builder: Standard Railway & Wagon Co Ltd   Lot No.: 3272
Diagram No.: 2/746   Built: 1960
Tare Weight: 20.5t + 21.5t   G.L.W.: 41.0t + 42.0t
Design Code: YV034A   YV034B +   Tops Code: YVP

ADB900810+   ADB900811

Number Series: B900906 - B900907

Description: 25T Bogie Weltrol Wagon
Builder: BR (Swindon Works)   Lot No.: 2879
Diagram No.: 2/754   Built: 1960
Tare Weight: 23.3t   G.L.W.: 48.8t
Design Code: YX069A   Tops Code: YXP

  B900907 (W)

Number Series: B900908 - B900910

Description: 25T Bogie Weltrol Wagon
Builder: BR (Swindon Works)   Lot No.: 2880
Diagram No.: 2/745   Built: 1959
Tare Weight: 18.5t   G.L.W.: 44.0t
Design Code: YV065A   Tops Code: YVP

  DB900910

Number Series: B900911

Description: 25T Bogie Weltrol Wagon
Builder: BR (Swindon Works)   Lot No.: 2977
Diagram No.: 2/739   Built: 1959
Tare Weight: 18.2t   G.L.W.: 43.7t
Design Code: YX058A   Tops Code: YXP

KDB900911

Number Series: B900912 - B900922

Description: 25T Bogie Weltrol Wagon
Builder: BR (Swindon Works)   Lot No.: 3102
Diagram No.: 2/741   Built: 1960
Tare Weight: 19.2t +   19.5t   G.L.W.: 44.7t +   45.0t
Design Code: YR002B *   YV016A = YV0046C   Tops Code: YVR +   YVP *
             YV046D +   YVQ   YRQ #

ADB900913*   ADB900914#   ADB900921+   ADB900922

Number Series: B900923 - B900924

```
Description: 25T Bogie Weltrol Wagon
Builder: Butterley Co Ltd              Lot No.: 3173
Diagram No.: 2/740                     Built: 1959
Tare Weight: 24.5t                     G.L.W.: 50.0t
Design Code: YV068A                    Tops Code: YVR

ADB900923 (W)
```

Number Series: B900925 - B900938

```
Description: 25T Bogie Weltrol Wagon
Builder: BR (Swindon Works)            Lot No.: 3192
Diagram No.: 2/741                     Built: 1960
Tare Weight: 19.2t *    19.5t          G.L.W.: 44.7t *    45.0t
Design Code: YR002C %  YV007B   YV016B *        Tops Code: YRR %   YVP
             YX040B #  YX040C >  YX040D <                  YVR *  YXP ><
                                                          YXR #

ADB900925*    ADB900928*   ADB900932*   KDB900935    KDB900937<
ADB900926*(W) ADB900929*   ADB900934*   KDB900936>   ADB900938*
 DB900927#    ADB900931%
```

Number Series: B901000 - B901005

```
Description: 40T Bogie Flatrol Wagon
Builder: BR (Swindon Works)            Lot No.: 2078
Diagram No.: 2/513                     Built: 1952
Tare Weight: 15.5t                     G.L.W.: 41.0t
Design Code: YV045C                    Tops Code: YVR

ADB901000      ADB901002
```

Number Series: B901014 - B901017

```
Description: 40T Bogie Flatrol Wagon
Builder: BR (Swindon Works)            Lot No.: 3193
Diagram No.: 2/748                     Built: 1960
Tare Weight: 21.5t                     G.L.W.: 61.5t
Design Code: YV017B                    Tops Code: YVR

ADB901014      ADB901015      ADB901016      ADB901017
```

Number Series: B901018 - B901021

```
Description: 40T Bogie Flatrol Wagon
Builder: BR (Swindon Works)            Lot No.: 3194
Diagram No.: 2/748                     Built: 1961
Tare Weight: 21.5t                     G.L.W.: 61.5t
Design Code: YV017B                    Tops Code: YVR

ADB901018      ADB901019      ADB901021
```

Number Series: B901022 - B901025

Description: 40T Bogie Flatrol Wagon
Builder: BR (Lancing Works)   Lot No.: 3241
Diagram No.: 2/731   Built: 1960
Tare Weight: 21.5t +   22.5t   G.L.W.: 62.0t +   63.0t
Design Code: YR003A   YR003B +   Tops Code: YRR   YRP +

ADB901022   ADB901023   ADB901025+

Number Series: B901102 - B901103

Description: 35T Bogie Weltrol Wagon
Builder: BR (Swindon Works)   Lot No.: 2647
Diagram No.: 2/734   Built: 1957
Tare Weight: 27.0t   G.L.W.: 62.5t
Design Code: YV014A   Tops Code: YVR

ADB901103

Number Series: B901104 - B901105

Description: 35T Bogie Weltrol Wagon
Builder: BR (Swindon Works)   Lot No.: 2881
Diagram No.: 2/751   Built: 1958
Tare Weight: 26.2t   G.L.W.: 61.7t
Design Code: YX071A   Tops Code: YXR

KDB901104

Number Series: B901106 - B901107

Description: 35T Bogie Weltrol Wagon
Builder: BR (Swindon Works)   Lot No.: 2978
Diagram No.: 2/747   Built: 1960
Tare Weight: 25.5t   G.L.W.: 61.0t
Design Code: YV064B   Tops Code: YVP

 DB901106

Number Series: B901152

Description: 50T Bogie Weltrol Wagon
Builder: BR (Swindon Works)   Lot No.: 3213
Diagram No.: 2/742   Built: 1960
Tare Weight: 24.5t   G.L.W.: 75.5t
Design Code: YR006A   Tops Code: YRP

ADB901152

Number Series: B901200 - B901203

Description: 55T Bogie Weltrol Wagon
Builder: BR (Ashford Works)   Lot No.: 2582
Diagram No.: 2/750   Built: 1964
Tare Weight: 48.5t   G.L.W.: 104.0t
Design Code: YV072A   Tops Code: YVW

ADB901200   ADB901201 (W)

Number Series: B901204

Description: 55T Bogie Weltrol Wagon
Builder: BR (Ashford Works)   Lot No.: 3016
Diagram No.: 2/750   Built: 1965
Tare Weight: 48.5t   G.L.W.: 104.0t
Design Code: YV072A   Tops Code: YVW

ADB901204

Number Series: B901205

Description: 55T Bogie Weltrol Wagon
Builder: BR (Ashford Works)   Lot No.: 3273
Diagram No.: 2/750   Built: 1965
Tare Weight: 48.5t   G.L.W.: 104.0t
Design Code: YV072A   Tops Code: YVW

ADB901205

Number Series: B901453

Description: 40T Bogie Trolley Wagon (Protrol EF)
Builder: BR (Ashford Works)   Lot No.: 3111
Diagram No.: 2/663   Built: 1963
Tare Weight: 22.5t   G.L.W.: 63.0t
Design Code: YX064A   Tops Code: YXP

KDB901453

Number Series: B901542 - B901576

Description: 40T Bogie Trolley Wagon
Builder: BR (Ashford Works)   Lot No.: 3247
Diagram No.: 2/682   Built: 1959
Tarte Weight: 28.5t   G.L.W.: 69.0t
Design Code: YV056C   Tops Code: YVR

DB901555 (W)   DB901557

Number Series: B901600 - B901603

Description: 55T Bogie Trestle Trolley Wagon
Builder: Teesside Bridge Engineering Ltd
Diagram No.: 2/681
Tare Weight: 36.0t
Design Code: YX018B

Lot No.: 2175
Built: 1950
G.L.W.: 92.0t
Tops Code: YXO

KDB901601 (W)

Number Series: B901702 - B901751

Description: 40T Bogie Trolley Wagon
Builder: BR (Derby Works)
Diagram No.: 2/682
Tare Weight: 27.5t =   28.5t
Design Code: YV056C   YV056E +   YX065A =

Lot No.: 3309
Built: 1960-61
G.L.W.: 68.0t =   69.0t
Tops Code: YVR   YVR +
                  YXQ =

ADB901702+   ADB901717   ADB901734=(W) ADB901744 (W)

Number Series: B901800 - B901801

Description: 135T Transformer Trolley Wagon
Builder: Head Wrightson & Co Ltd
Diagram No.: 2/470
Tare Weight: 73.10t
Design Code: FR005A

Lot No.: 2419
Built: 1953
G.L.W.: 208.0t
Tops Code: FRO

 B901801 (W)

Number Series: B902601 - B902619

Description: Barrier Brake Van
Builder: BR (Stewarts Lane)
Diagram No.: 1/816
Tare Weight: 17.6t
Design Code: RF013A

Lot No.: 4041
Built: 1984
G.L.W.: 17.6t
Tops Code: RFQ

(Former numbers in brackets)

B902601 (W) (S94643)	B902608 (S94223)	B902612 (W) (E94477)
B902606 (W) (M94736)	B902609 (W) (M94169)	B902614 (W) (S94893)
B902607 (M94897)	B902611 (W) (S94345)	

Number Series: B902805 - B902810

Description: 290T Boiler Wagon (4-part set)
Builder: BR (Ashford Works)
Diagram No.: 2/034
Tare Weight: 32.5t
Design Code: YV069A   YV069B +

Lot No.: 3527
Built: 1965
G.L.W.: 108.5t
Tops Code: YVV   YVP +

 DB902805   DB902806   DB902807+   DB902808+

Number Series: B904082 - B904101

Description: 20T Lowmac Wagon
Builder: BR (Swindon Works)   Lot No.: 2324
Diagram No.: 2/245   Built: 1953
Tare Weight: 12.5t   G.L.W.: 33.0t
Design Code: ZV027A + ZV027C   Tops Code: ZVO + ZVR

ADB904086+   ADB904092

Number Series: B904102 - B904104

Description: 20T Lowmac Wagon
Builder: BR (Swindon Works)   Lot No.: 2492
Diagram No.: 2/245   Built: 1954
Tare Weight: 11.9t   13.0t +   G.L.W.: 32.4t   33.0t +
Design Code: ZV027D   ZX026B +   Tops Code: ZVP + ZVW

ADB904102+   ADB904103

Number Series: B904105 - B904119

Description: 20T Lowmac Wagon
Builder: BR (Swindon Works)   Lot No.: 2592
Diagram No.: 2/245   Built: 1955
Tare Weight: 11.9t =   12.5t   G.L.W.: 32.4t =   33.0t
Design Code: ZV027D = ZX053A ZY096B +   Tops Code: ZVW = ZXV
                                                          ZYW +

   DB904107 (W)   DB904111   LDB904112+   ADB904115=

Number Series: B904120 - B904122

Description: 20T Lowmac Wagon
Builder: Teesside Bridge Eng Co Ltd   Lot No.: 2959
Diagram No.: 2/249   Built: 1957
Tare Weight: 12.5t   G.L.W.: 35.0t
Design Code: ZR202A   Tops Code: ZRV

ADB904122

Number Series: B904123 - B904134

Description: 20T Lowmac Wagon
Builder: BR (Swindon Works)   Lot No.: 2976
Diagram No.: 2/250   Built: 1958
Tare Weight: 12.5t   G.L.W.: 33.0t
Design Code: ZX155A   ZR190A *   Tops Code: ZRV + ZXV

   DB904129   ADB904131*(W)   DB904132   DB904133

Number Series: B904135 - B904144

Description: 20T Lowmac Wagon
Builder: BR (Swindon Works)
Diagram No.: 2/752
Tare Weight: 12.5t
Design Code: ZR191A * ZV216A = ZV216B

Lot No.: 3101
Built: 1958
G.L.W.: 12.5t 33.0t *=
Tops Code: ZRV * ZVV =
ZVW

ADB904135* ADB904140 (W) ADB904141 ADB904142 ADB904144=
ADB904136

Number Series: B904145 - B904154

Description: 21T Lowmac Wagon
Builder: BR (Swindon Works)
Diagram No.: 2/240
Tare Weight: 11.0t
Design Code: ZV070D + ZV070G : ZX057A *
ZX057B

Lot No.: 3254
Built: 1960
G.L.W.: 32.5t
Tops Code: ZVP + ZVR :
ZXP

KDB904145 (W) ADB904148+(W) ADB904149+(W) ADB904150:(W) KDB904152*
KDB904147

Number Series: B904155 - B904157

Description: 20T Lowmac Wagon
Builder: BR (Swindon Works)
Diagram No.: 2/252
Tare Weight: 12.5t
Design Code: ZV216A

Lot No.: 3298
Built: 1961
G.L.W.: 33.0t
Tops Code: ZVV

ADB904155 (W)

Number Series: B904500 - B904537

Description: 25T Lowmac Wagon
Builder: P & W McLellan Ltd
Diagram No.: 2/242
Tare Weight: 13.5t 14.0t -#+ 14.4t %:
Design Code: ZV187B % ZV187C : ZX029B # ZX029C + ZX050A
ZX050B = ZY111A -
Tops Code: ZVV % ZVW : ZXV # ZXV ZXW += ZYW -

Lot No.: 2187
Built: 1950
G.L.W.: 29.0t #+ 39.0t
39.5t -%:

DB904500+   DB904508#   LDB904514-   LDB904523-   TDB904529%(W)
DB904502   DB904510   DB904518=   DB904524=   ADB904532:
DB904503 (W) DB904512 (W) DB904519=   DB904525=   ADB904534:
DB904505 (W) DB904513=   DB904520

Number Series: B904538 - B904565

Description: 25T Lowmac Wagon
Builder: BR (Swindon Works)
Diagram No.: 2/243
Tare Weight: 13.5t
Design Code: ZV005B   ZX051A +   ZV025E *

Lot No.: 2268
Built: 1952
G.L.W.: 39.0t
Tops Code: ZVP   ZVP *
                ZXO +

ADB904541 (W) ADB904542*(W)   DB904547+(W)

Number Series: B904572 - B904629

Description: 25T Lowmac Wagon
Builder: BR (Swindon Works)
Diagram No.: 2/243
Tare Weight: 13.1t
Design Code: ZV025F

Lot No.: 2593
Built: 1954
G.L.W.: 39.0t
Tops Code: ZVR

ADB904585

Number Series: B904630 - B904633

Description: 25T Lowmac Wagon
Builder: BR (Swindon Works)
Diagram No.: 2/246
Tare Weight: 13.5t
Design Code: ZX054A   ZX054B +

Lot No.: 2594
Built: 1955
G.L.W.: 39.0t
Tops Code: ZXV   ZXW +

DB904630 (W)   DB904631+   DB904632 (W) DB904633+(W)

Number Series: B904634 - B904641

Description: 25T Lowmac Wagon
Builder: Derbyshire C & W Co Ltd
Diagram No.: 2/254
Tare Weight: 13.9t + 14.0t
Design Code: ZV207A   ZV215A +

Lot No.: 2645
Built: 1956
G.L.W.: 39.0t   39.4t +
Tops Code: ZVW

ADB904638+   ADB904639+   ADB904640

Number Series: B904662 - B904674

Description: 25T Lowmac Wagon
Builder: Derbyshire C & W Co Ltd
Diagram No.: 2/247
Tare Weight: 13.5t   14.0t =
Design Code: ZV028A   ZV028B =

Lot No.: 2714
Built: 1955
G.L.W.: 39.0t
Tops Code: ZVR   ZVQ =

ADB904662=   ADB904663=   DB904667   ADB904668 (W) ADB904671

Number Series: B904695 - B904709

Description: 25T Lowmac Wagon
Builder: Derbyshire C & W Co Ltd
Diagram No.: 2/253
Tare Weight: 13.9t   14.0t %   14.1t :
Design Code: ZX029D : ZX097A % ZX151A +
             ZX151B

Lot No.: 3172
Built: 1958
G.L.W.: 36.6t   39.4t +
             39.5t %
Tops Code: ZXA :   ZXV %+
             ZXW

```
   DB904696        RDB904698%(W) TDB904706+   TDB904708:      DB904709
   DB904697        DB904700
```

Number Series: B904710 - B904722

Description: 25T Lowmac Wagon
Builder: BR (Swindon Works)
Diagram No.: 2/253
Tare Weight: 13.9t   14.0t +
Design Code: ZV020E + ZX151A

Lot No.: 3198
Built: 1960
G.L.W.: 36.5t + 39.4t
Tops Code: ZVW +   ZXV

```
   DB904710        DB904716        ADB904719+
```

Number Series: B905000 - B905019

Description: 15T Lowmac Wagon
Builder: BR (Swindon Works)
Diagram No.: 2/241
Tare Weight: 9.2t   10.0t +
Design Code: ZV024C   ZX048D +

Lot No.: 2087
Built: 1950
G.L.W.: 24.2t 25.0t +
Tops Code: ZVR   ZXR +

```
ADB905000 (W)   DB905004+
```

Number Series: B905020 - B905059

Description: 14T Lowmac Wagon
Builder: BR (Shildon Works)
Diagram No.: 2/244
Tare Weight: 8.5t
Design Code: ZY040C   ZY040D +

Lot No.: 2475
Built: 1953
G.L.W.: 22.5t
Tops Code: ZYR ZYW +

```
LDB905034+     LDB905052 (W) LDB905053    LDB905056 (W)
```

Number Series: B905060 - B905071

Description: 14T Lowmac Wagon
Builder: BR (Shildon Works)
Diagram No.: 2/244
Tare Weight: 8.5t
Design Code: ZY040C

Lot No.: 2553
Built: 1954
G.L.W.: 22.5t
Tops Code: ZYR

```
LDB905062 (W) LDB905065 (W)
```

Number Series: B905072 - B905077

Description: 15T Lowmac Wagon
Builder: BR (Swindon Works)   Lot No.: 2591
Diagram No.: 2/248   Built: 1956
Tare Weight: 9.7t   G.L.W.: 24.7t
Design Code: ZV029F   Tops Code: ZVW

ADB905076 (W)

Number Series: B905078 - B905087

Description: 15T Lowmac Wagon
Builder: BR (Swindon Works)   Lot No.: 2875
Diagram No.: 2/248   Built: 1956
Tare Weight: 9.7t + 10.0t   G.L.W.: 24.7t + 25.0t
Design Code: ZV153A # ZX055A   ZX055G :   Tops Code: ZVV # ZXV
             ZY125A +   ZYW +

KDB905078 (W) LDB905080+(W) KDB905083:   KDB905086#

Number Series: B905088 - B905097

Description: 15T Lowmac Wagon
Builder: BR (Swindon Works)   Lot No.: 2975
Diagram No.: 2/241   Built: 1957
Tare Weight: 9.3t # 10.0t   20.0t +   G.L.W.: 24.3t #   25.0t
Design Code: ZX027D + ZX048B   ZX048C #   Tops Code: ZXA # ZXR +
                                                          ZXP

 DB905089   DB905094 (W) DB905095   KDB905096+   DB905097
TDB905093#

Number Series: B905098 - B905116

Description: 15T Lowmac Wagon
Builder: BR (Swindon Works)   Lot No.: 3100
Diagram No.: 2/348   Built: 1958
Tare Weight: 9.7t * 10.0t   G.L.W.: 25.0t
Design Code: ZX055B   ZX055D * ZX055E =   Tops Code: ZXV   ZXW =*
             ZY114A -   ZYW -

KDB905098   KDB905102=   ADB905109+(W) KDB905111=   KDB905114
KDB905100 (W) LDB905108-   KDB905110*(W) KDB905112   KDB905115 (W)

Number Series: B909000 - B909020

Description: 38T Rectank Wagon
Builder: Butterley Wagon Co Ltd   Lot No.: 3165
Diagram No.: 2/440   Built: 1960
Tare Weight: 10.0t +   15.4t   G.L.W.: 32.5t + 53.9t
Design Code: YV044C + YX003A   Tops Code: YVO + YXP

KDB909004   ADB909011+

Number Series: B909048 - B909077

Description: 38T Rectank Wagon
Builder: BR (Swindon Works)   Lot No.: 3299
Diagram No.: 2/440   Built: 1960
Tare Weight: 15.5t +   16.5t   G.L.W.: 55.0t
Design Code: YR001B   YX003C +   YR001D #   Tops Code: YRR   YXR +

ADB909053 (W) ADB909059#(W) KDB909064+   KDB909074+

Number Series: B911000 - B911999

Description: 13T Bolster Wagon
Builder: BR (Shildon Works)   Lot No.: 2783
Diagram No.: 1/400   Built: 1956
Tare Weight: 9.0t   G.L.W.: 22.0t
Design Code: ZS011C   Tops Code: ZSR

  DB911222 (W)   DB911261

Number Series: B914000 - B915999

Description: 13T Bolster Wagon
Builder: BR (Shildon Works)   Lot No.: 3005
Diagram No.: 1/402   Built: 1957
Tare Weight: 9.0t   G.L.W.: 9.0t
Design Code: ZS012A +   ZS012B :   ZS012C   Tops Code: ZSO +   ZSP :
                                                         ZSR

 DB914152   DB914415+(W)   DB914750 (W)   DB915342   DB915469
 DB914240+(W)   DB914534   DB915123:   DB915352 (W)   DB915763
 DB914378+(W+   DB914629   DB915198   DB915355   DB915774
 DB914383:   DB914728   ADB915318:

Number Series: B916000 - B917999

Description: 13T Bolster Wagon
Builder: BR (Shildon Works)   Lot No.: 3006
Diagram No.: 1/402   Built: 1957
Tare Weight: 9.0t   G.L.W.: 22.0t
Design Code: ZS012A +   ZS012B :   ZS012C   Tops Code: ZSO +   ZSP :
                                                         ZSR

ADB916139+(W)   DB916608   DB916738 (W)   DB917143   DB917363+(W)
 DB916452:   DB916630   DB916742 (W)   DB917212   DB917737
 DB916549+   DB916719   DB917090   DB917309 (W)

<u>Number Series: B920000 - B920199</u>

Description: 21T Double Bolster Wagon
Builder: BR (Wolverton Works)   Lot No.: 2020
Diagram No.: 1/416   Built: 1949
Tare Weight: 10.0t   10.5t +   G.L.W.: 32.0t
Design Code: ZD085A ZN002B +   Tops Code: ZDO

  DB920093+(W) TDB920124 (W) DB920197+(W)

<u>Number Series: B920200 - B920399</u>

Description: 21T Double Bolster Wagon
Builder: BR (Shildon Works)   Lot No.: 2035
Diagram No.: 1/415   Built: 1949
Tare Weight: 10.5t   G.L.W.: 32.0t
Design Code: ZN001A   Tops Code: ZNO

  DB920217 (W)   DB920252 (W)   DB920311 (W)   DB920321 (W)

<u>Number Series: B922500 - B922699</u>

Description: 30T Bogie Bolster Wagon   Lot No.: 3162
Builder: Metropolitan Cammell C & W Co Ltd   Built: 1958
Diagram No.: 1/477   G.L.W.: 18.0t #   36.6t :
Tare Weight: 13.6t :   14.5t >   18.0t   45.0t >   48.5t
Design Code: BC003C + YN048A = YN048B < YS031A * YS034B #
             YV057C " YV057E > YV057F % YV057L   YY012E :
Tops Code: BCV + YNV = YNW < YSV # YSW * YVV "> YVW   YYW :

   B922515+(W) LDB922544:(W) B922616+   DB922648#   KDB922678%
  KDB922520   LDB922572:   B922618+(W) ADB922649>   KDB922687"(W)
  LDB922530:(W) KDB922582=(W) B922622+(W) KDB922656=(W) KDB922694"
   B922536+(W) B922604+(W) DB922640#   KDB922673<

<u>Number Series: B922700 - B922999</u>

Description: 30T Bogie Bolster Wagon   Lot No.: 3200
Builder: BR (Swindon Works)   Built: 1959
Diagram No.: 1/477   G.L.W.: 18.0t :   36.6t #
Tare Weight: 13.6t #   17.5t %<   18.0t   48.5t
Design Code: BC003C + YN048A > YS034B : YV057C " YV057F %
             YV057J * YX035A ! YX035B < YY012E #
Tops Code: BCV + YYW # YNV >< YSV : YVV " YVW *% YXV !

KDB922713"   DB922757:   KDB922842>(W) KDB922946!   B922975+(W)
LDB922720#   KDB922766%   KDB922855%(W) KDB922952>(W) LDB922985#(W)
  B922723+(W) KDB922770"   LDB922885#   KDB922953%   LDB922990#(W)
LDB922737#   KDB922799*(W)   B922891+(W) KDB922954>   B922996+(W)
 DB922744+(W) KDB922802>(W) LDB922912#(W) KDB922955%   KDB922997%
 DB922752+(W) KDB922809%   DB922925<

89

Number Series: B923000 - B923099

Description: 30T Bogie Bolster Wagon   Lot No.: 3238
Builder: BR (Swindon Works)   Built: 1959
Diagram No.: 1/477   G.L.W.: 18.0t - 36.6t !
Tare Weight: 13.6t !   17.5t >#   18.0t   48.5t
Design Code: BC003C + YN048A % YS031A * YS034B - YV057C <
             YV057F > YV057H = YV057K # YY012E !
Tops Code: BCV + YNV % YYW ! YSV - YSW * YVV < YVW >=#

KDB923000=   LDB923018!   KDB923068>   KDB923078<(W) KDB923095<
 DB923005+(W) KDB923019%(W) DB923074%(W)   B923081+(W) KDB923096%(W)
LDB923010!(W) DB923037-(W) KDB923076>   LDB923083!   DB923099*
LDB923014!(W) KDB923040#

Number Series: B923100 - B923299

Description: 30T Bogie Bolster Wagon
Builder: BR (Swindon Works)   Lot No.: 3341
Diagram No.: 1/477   Built: 1961
Tare Weight: 15.65t   16.1t >+   16.25t ! G.L.W.: 17.7t #= 46.5t !
             17.5t >#   17.7t #=   18.0t <*   47.1t
             18.2t -   48.5t +-><*:%
Design Code: BC003E - BT001C + BT002C > BQ001E ! BQ001H   YN048A <
             YN048C * YV057C % YV057F : YX035D # YS034A =
Tops Code: BCW - BTW +> BQV   BQW ! YNV < YNW * YSW = YVV %
           YVW :   YXW #

   B923100+(W)   B923139+(W)   B923186!(W)   B923224   KDB923269*
KDB923103*(W) KDB923140*(W)   B923187+(W)   B923225 (W) KDB923270<(W)
   B923104 (W)   B923142 (W) KDB923188*(W)   B923226 (W) KDB923271+(W)
   B923106   B923144>   KDB923189<   B923230   B923272>(W)
KDB923107*   B923145 (W)   B923190 (W)   B923232   B923273
   B923108 (W)   B923146   KDB923191<   B923233 (W)   B923274 (W)
KDB923109:(W) KDB923147<   B923192   B923234 (W) KDB923275*
KDB923110*(W)   B923148>(W) KDB923194*(W) KDB923235-   B923276
   B923111-(W)   B923149+(W)   B923196-(W)   B923240 (W)   B923277=(W)
KDB923115*(W)   B923152+(W)   B923200 (W)   B923242 (W) KDB923278-(W)
   B923117   B923155   B923201   B923245 (W)   B923279>(W)
KDB923118%   DB923157#   B923203   B923246+(W) KDB923280*(W)
   B923119-(W)   B923159 (W)   B923204+   B923247-(W)   B923284
   B923120>(W)   B923162 (W)   B923205 (W)   B923248+(W)   B923285
   B923122+(W)   B923164>(W)   B923206 (W)   B923249   B923286
KDB923123*   B923165>(W)   B923209>   B923250   B923288
   B923124 (W)   B923167>(W)   B923210   KDB923251-   B923289
   B923125>(W)   B923168>(W) KDB923211*   B923253 (W)   B923290
   B923126>(W) KDB923171*(W)   B923212 (W)   B923254+(W) KDB923291%(W)
   B923131   B923172   B923214   B923255-(W) KDB923293:(W)
   B923132+   B923173 (W)   B923218   B923259 (W) KDB923294>(W)
   B923133+(W)   B923174   B923219>   B923262>   KDB923296*(W)
   B923134 (W)   B923175 (W) KDB923220:   KDB923264<   B923297 (W)
   B923136+(W) KDB923176%   B923221   B923265>(W)   B923298 (W)
   B923137   B923185 (W)   B923223 (W)   B923266   B923299-(W)
   B923138>(W)

Number Series: B923300 - B924399

Description: 30T Bogie Bolster Wagon   Lot No.: 3343
Builder: BR (Ashford Works)   Built: 1961
Diagram No.: 1/479   G.L.W.: 47.5t >= 48.4t :
Tare Weight: 15.0t >   15.9t :   16.0t   48.5t
Design Code: YE003A   YN047A +   YN047B -   YN047C !   YN047D "
            YN059A %   YR008A *   YV054A :   YV058A =   YV058B >
            YY035A #   YY035B ^   YY035D <
Tops Code: YEV   YNV +!-%   YNW "   YRV *   YVV :=>   YYV #   YYW ^<

```
DB923305+(W)   LDB923552#(W)  LDB923745#(W)   DB923994+(W)  KDB924207=(W)
KDB923319:(W)  KDB923559*(W)  LDB923786>     KDB923998:(W)   DB924236+(W)
KDB923323:(W)  KDB923563:(W)  LDB923792#(W)  LDB924062#      DB924257+(W)
KDB923331:(W)  KDB923569:(W)  LDB923797^      DB924075+(W)  KDB924265-
KDB923365*(W)   DB923589+(W)   DB923798+(W)   DB924077       KDB924267:(W)
 DB923378+(W)  KDB923599*(W)  KDB923805>(W)  KDB924080:(W)    DB924298+(W)
KDB923437:(W)  LDB923603#(W)   DB923819+(W)  KDB924085>(W)  LDB924299#(W)
KDB923443:(W)  KDB923607%(W)  KDB923821:(W)   DB924086       KDB924310:(W)
 DB923474+(W)   DB923610+(W)  LDB923840^     KDB924093:(W)  KDB924316*(W)
 DB923477+(W)  KDB923617>(W)  KDB923857*(W)  KDB924105:(W)    DB924326+(W)
 DB923481+(W)   DB923671+(W)   DB923880+(W)  KDB924108=(W)  KDB924347=(W)
KDB923485*(W)   DB923674+       DB923883+(W)   DB924114+(W)  KDB924373=(W)
 DB923487+(W)  KDB923686>(W)  LDB923903#(W)  KDB924136=(W)  KDB924382=(W)
KDB923523+(W)   DB923693+(W)  KDB923932*(W)  KDB924147!(W)    DB924383+(W)
KDB923503!(W)  KDB923694*(W)  KDB923934*(W)   DB924163+(W)    DB924386+(W)
 DB923510+(W)   DB923701+(W)  KDB923936:(W)  KDB924170!(W)    DB924391+(W)
 DB923523+(W)  LDB923703#(W)   DB923954+(W)  LDB924176#      LDB924396^
KDB923528!(W)  KDB923706!(W)  KDB923962"      DB924177+(W)  KDB924399*(W)
 DB923536+(W)   DB923718+(W)   DB923984+(W)  KDB924203"
```

Number Series: B924400 - B924799

Description: 30T Bogie Bolster Wagon   Lot No.: 3397
Builder: Metropolitan Cammell Co Ltd   Built: 1961
Diagram No.: 1/477   G.L.W.: 17.7t *y 45.0t <
Tare Weight: 15.65t   16.1t "#   16.25t %   46.5t %   47.1t
            17.5t ^!   17.7t *y 17.8t <   48.5t >"#+z:x^!
            18.0t xz: 18.2t >+
Design Code: BC003E >   BT001C "   BT002C #   BQ001E %   BQ001H   RS005A +
            YN048A z   YN048C :   YS034A *   YV057C x   YV057F ^
            YV057K !   YX035D y   YX035E <
Tops Code: BCW >   BTW "   BQW   RSW +   YNV z   YNW :   YSV *   YVV x
          YVW !   YXW y   YXV <

```
KDB924401:(W)   B924424        B924444"(W)   B924471        DB924491*
 B924402        B924426 (W)    B924445 (W)   B924472        DB924492:
KDB924408:      B924427 (W)    B924446 (W)   B924473         B924493#(W)
 B924410 (W)    B924428 (W)  KDB924447:      B924474 (W)     B924494 (W)
 B924412 (W)  KDB924429<      B924452#(W)    B924475#(W)     B924497"
 B924413 (W)    B924432 (W)    B924454 (W)   B924476#(W)     B924501 (W)
 B924414 (W)    B924433#     KDB924455:     KDB924478:(W)    B924505#
KDB924416:     KDB924436:(W)   B924456 (W)   B924481 (W)     B924508#(W)
 B924418 (W)    B924437 (W)    B924457#(W)  KDB924482:(W)    B924511
 B924419#       B924439 (W)    B924458#(W)   B924484 (W)     B924512+(W)
 B924420 (W)    B924440 (W)    B924459        B924485 (W)  KDB924513:(W)
 B924421"(W)    B924441 (W)    B924461 (W)  KDB924486:       B924514>(W)
 B924422 (W)  KDB924442:(W)    B924469>(W)   B924487 (W)     B924515>(W)
 B924423        B924443>(W)    B924470"(W)   B924490 (W)     B924516
```

```
B924520 (W)      B924570 (W)      B924627         KDB924689:(W)    B924753 (W)
B924522>(W)      B924571 (W)      KDB924628:(W)    B924693 (W)      B924755>(W)
B924523%(W)      B924572 (W)      KDB924629 (W)    B924694#         B924756#
KDB924526:       B924573#(W)      KDB924630:(W)    B924697 (W)      B924757
KDB924527:(W)    KDB924574:(W)    KDB924632:       B924698          B924760#(W)
B924528 (W)      B924575          B924633 (W)      B924701 (W)      B924763 (W)
B924530 (W)      B924578>         B924634"(W)      B924703          B924764
B924535#(W)      B924580 (W)      KDB924637:       KDB924704:(W)    B924765#(W)
B924536 (W)      KDB924581:(W)    B924638"(W)      B924708          KDB924767:(W)
B924538 (W)      B924582          B924640#         KDB924710:(W)    B924768 (W)
B924539 (W)      B924585#         B924643#(W)      B924711 (W)      B924770 (W)
DB924540^(W)     B924586>(W)      B924646          B924712          B924771<(W)
B924541#         B924589 (W)      B924647:(W)      B924714 (W)      KDB924772%
KDB924542x(W)    B924590*(W)      B924648>(W)      B924715#(W)      KDB924774:
KDB924544>(W)    B924598          KDB924649:       KDB924716z(W)    B924775>(W)
B924545#(W)      B924599          B924651          B924720>(W)      B924776#(W)
B924546>         DB924600*(W)     B924653"(W)      B924722 (W)      KDB924777:(W)
KDB924547!       B924601#         B924654#(W)      B924724>(W)      B924778"(W)
B924548 (W)      KDB924602<       KDB924656:       B924726#(W)      B924779 (W)
B924550          B924604"(W)      KDB924657:       B924727          B924780 (W)
B924551"(W)      B924605#(W)      B924659#(W)      B924731          B924781#(W)
B924552          B924606"(W)      KDB924663>(W)    B924732#         B924783>(W)
DB924553y        B924607          B924664          B924736>(W)      B924784 (W)
B924554          KDB924608:(W)    B924665"(W)      B924738          B924785#
B924555          KDB924609:       B924669"(W)      B924740 (W)      B924786 (W)
B924558 (W)      B924610 (W)      B924670"         B924741          B924787
B924560 (W)      B924612>(W)      KDB924671:(W)    B924742"(W)      B924788 (W)
B924561>(W)      B924614          KDB924672:(W)    B924743          B924789
B924563"(W)      B924615"(W)      B924673#(W)      DB924744y        B924790
KDB924564:(W)    B924616 (W)      KDB924674:(W)    KDB924747:(W)    KDB924793:
B924565#(W)      B924618>(W)      B924675          B924748          B924794"(W)
B924566          B924620 (W)      B924679 (W)      B924750          B924795>(W)
B924568#(W)      B924622#(W)      B924680>(W)      KDB924751:       B924796 (W)
KDB924569:(W)    KDB924623:(W)    B924687"(W)      KDB924752:(W)    B924799 (W)
```

Number Series: B924800 - B924899

Description: 30T Bogie Bolster Wagon   Lot No.: 3440
Builder: BR (Ashford Works)   Built: 1962
Diagram No.: 1/479   G.L.W.: 12.0t +   42.0t =
Tare Weight: 12.0t +=   16.0t   48.5t
Design Code: YN047A *   YY035A   ZC514A = Tops Code: YNV *   YYV
             ZS140A +   ZCA = ZSA +

```
DB924807+        DB924824=        DB924832*(W)     DB924848=        DB924860+
DB924819*(W)     LDB924826        DB924835*(W)     DB924857*(W)     DB924875*
DB924820=        DB924828*
```

Number Series: B927400 - B927499

Description: 42T Bogie Bolster Wagon   Lot No.: 3104
Builder: BR (Swindon Works)   Built: 1958
Diagram No.: 1/472   G.L.W.: 64.8t *   65.0t +
Tare Weight: 22.3t *   22.5t +   23.5t   66.0t
Design Code: YN041C =   YN041D   YY039B + Tops Code: YNV =   YNW
             YY039C *   YYR *   YYW +

```
LDB927400+(W) LDB927433+     ADB927473=(W) LDB927484+     LDB927496+
LDB927403+     LDB927451*(W) KDB927482 (W) KDB927486 (W)
```

Number Series: B927500 - B927599

```
Description: 42T Bogie Bolster Wagon      Lot No.: 3113
Builder: BR (Lancing Works)               Built: 1958
Diagram No.: 1/472                        G.L.W.: 66.0t
Tare Weight: 23.5t                        Tops Code: YNV    YNW *
Design Code: YN041C   YN041D * YY039A +              YYV +
```

```
LDB927519+    KDB927522*    KDB927532*    KDB927541 (W) KDB927549 (W)
```

Number Series: B927600 - B927799

```
Description: 42T Bogie Bolster Wagon      Lot No.: 3246
Builder: BR (Lancing Works)               Built: 1959
Diagram No.: 1/478                        G.L.W.: 22.4t * 22.5t +
Tare Weight: 22.4t *   22.5t                       65.0t
Design Code: YN050A = YN050B > YN050C : YS037A * YY038A # YY038B
Tops Code: YNV = YNW >: YSV * YSW + YYV # YYW
```

```
LDB927644#(W) LDB927679      LDB927712     KDB927735:(W) KDB927769:(W)
KDB927648:(W) LDB927681      LDB927718#    LDB927736#    LDB927779 (W)
KDB927649:(W) KDB927692:(W) ADB927727=(W) ADB927737>(W) LDB927780 (W)
ADB927657*(W) LDB927704 (W) ADB927730*    KDB927753:(W) ADB927781=(W)
LDB927659     LDB927707 (W) LDB927732     KDB927760:(W) LDB927785 (W)
LDB927672
```

Number Series: B927800 - B927999

```
Description: 42T Bogie Bolster Wagon
Builder: Powell Duffryn Co Ltd            Lot No.: 3407
Diagram No.: 1/484                        Built: 1961
Tare Weight: 22.4t :!   22.5t             G.L.W.: 65.0t
Design Code: BS001A ! BD003D : YN051A *   Tops Code: BDW : BSW !
             YN051B   YN051D % YS029A +              YNV * YNW
             YY042B #                               YSW + YYW #
```

```
KDB927800%    KDB927836 (W) KDB927861      KDB927892%(W) KDB927913
KDB927801 (W)  B927837:(W) KDB927864 (W) KDB927893 (W)  B927915:(W)
KDB927806 (W) KDB927838 (W) KDB927866 (W)  B927894:(W) KDB927916%(W)
KDB927807 (W) KDB927839     KDB927867 (W) KDB927896     KDB927918
KDB927808 (W)  B927840:(W) KDB927870 (W) KDB927897 (W) KDB927924 (W)
KDB927809%(W) LDB927842#    KDB927871 (W)  B927898:(W) KDB927927 (W)
KDB927810%(W) KDB927844     KDB927872 (W) KDB927899 (W) KDB927928
KDB927812 (W) KDB927847%(W) KDB927876      B927900:(W) KDB927929 (W)
KDB927816     KDB927848 (W) KDB927877     KDB927901     KDB927930 (W)
KDB927818 (W) KDB927849     KDB927879 (W) KDB927904      B927932:(W)
KDB927821 (W)  B927853:(W) KDB927880     KDB927907 (W) KDB927934 (W)
KDB927825 (W) KDB927854     KDB927882 (W) KDB927908 (W)  B927935:(W)
ADB927826*(W) KDB927856 (W) KDB927886 (W) KDB927909 (W)  B927936:(W)
KDB927829 (W)  B927857:(W) KDB927887 (W) KDB927910 (W) KDB927937 (W)
 B927832:(W) KDB927858     KDB927889 (W) KDB927911 (W) KDB927939 (W)
```

```
KDB927940 (W)  KDB927949+       B927963:(W)  KDB927976       KDB927990 (W)
KDB927941      KDB927951 (W)  KDB927964 (W)  KDB927977 (W)  KDB927993 (W)
KDB927942 (W)    B927952:(W)  KDB927965 (W)    B927979:(W)    B927994:(W)
KDB927944      KDB927955 (W)  KDB927966%(W)  KDB927982 (W)    B927997:(W)
KDB927945 (W)  KDB927957 (W)  KDB927967 (W)    B927983:(W)  KDB927998 (W)
KDB927946 (W)  KDB927960 (W)  KDB927968      KDB927985 (W)  KDB927999 (W)
KDB927948 (W)  KDB927962 (W)  KDB927969 (W)  KDB927989 (W)
```

Number Series: B928000 - B929199

```
Description: 42T Bogie Bolster Wagon
Builder: Charles Roberts & Co Ltd        Lot No.: 3408
Diagram No.: 1/484                       Built: 1962
Tare Weight: 22.4t :#  22.5t             G.L.W.: 65.0t
Design Code: BD003D #  BS001B :  YN051B   Tops Code: BDW #  BSR :
             YN051C *  YN051D !  YY042B +            YNW     YYW +
```

```
LDB928000+(W)  KDB928038 (W)    B928074#(W)  KDB928115*(W)  KDB928156
KDB928001 (W)  KDB928040 (W)  KDB928075      ADB928119      KDB928157 (W)
  B928003#(W)  KDB928042      KDB928076      KDB928120 (W)  KDB928158
KDB928004      KDB928043 (W)  KDB928077        B928121#(W)  KDB928159
KDB928006      KDB928044      KDB928082 (W)    B928122#(W)  KDB928161 (W)
  B928007#(W)    B928046#(W)  KDB928083 (W)    B928125#(W)  KDB928164
  B928008#(W)  KDB928047 (W)  KDB928084 (W)    B928128#(W)  KDB928165
KDB928009 (W)  KDB928050 (W)  KDB928085      KDB928129 (W)  KDB928170 (W)
KDB928010      LDB928052+       B928086#(W)    B928132:(W)  KDB928172
KDB928011      KDB928054 (W)  KDB928087        B928133 (W)  KDB928173
KDB928013      KDB928055      KDB928089        B928134#(W)  KDB928174
  B928014#(W)    B928056#(W)  KDB928091      LDB928135+     KDB928175!(W)
KDB928018 (W)  KDB928057 (W)  KDB928094 ~(W)    B928137#(W)  KDB928177!(W)
KDB928021      KDB928058 (W)  KDB928096      KDB928139      KDB928178
  B928023#(W)  KDB928059 (W)  KDB928097        B928142#(W)  KDB928179 (W)
KDB928026 (W)  KDB928060 (W)  KDB928100 (W)  KDB928143      KDB928181
KDB928027 (W)  KDB928061      KDB928102 (W)  KDB928145      KDB928184
KDB928029      KDB928063 (W)  KDB928103      KDB928147      KDB928185
KDB928030 (W)  KDB928067      KDB928106 (W)  KDB928148 (W)    B928186#(W)
  B928031:(W)  KDB928068      KDB928108 (W)  KDB928150      KDB928188 (W)
KDB928033 (W)  KDB928069      KDB928109 (W)  LDB928151+     KDB928193!(W)
KDB928034 (W)  KDB928070 (W)  KDB928111!(W)    B928152#(W)    B928194#(W)
KDB928037 (W)  KDB928073 (W)  KDB928112      KDB928153      KDB928196 (W)
```

Number Series: B930000 - B930249

```
Description: 22T Plate Wagon
Builder: BR (Shildon Works)              Lot No.: 2037
Diagram No.: 1/430                       Built: 1949
Tare Weight:  13.0t                      G.L.W.: 13.0t
Design Code: ZS013B                      Tops Code: ZSP
```

ADB930227

Number Series: B930550 - B931049

Description: 22T Plate Wagon
Builder: G R Turner Ltd   Lot No.: 2151
Diagram No.: 1/430   Built: 1950
Tare Weight: 13.0t   G.L.W.: 13.0t
Design Code: ZS013B + ZS013C   Tops Code: ZSP

  DB930855+(W)   DB930878 (W)

Number Series: B931050 - B931589

Description: 22T Plate Wagon   Lot No.: 2199
Builder: BR (Shildon Works)   Built: 1951
Diagram No.: 1/431   G.L.W.: 13.0t   32.5t *
Tare Weight: 10.0t *   11.0t :#   13.0t   33.5t :#
Design Code: ZS014C   ZS014E +   ZS097C :   Tops Code: ZSP   ZSV :
             ZV044D #   ZX129B *   ZVV #   ZXV *

  DB931139+(W)   DB931381:(W) KDB931486#(W) ADB931529   KDB931563*(W)

Number Series: B931590 - B931749

Description: 22T Plate Wagon
Builder: BR (Shildon Works)   Lot No.: 2327
Tare Weight: 9.8t   10.5t +   Built: 1952
Diagram No.: 1/431   G.L.W.: 9.8t   32.5t +
Design Code: ZD087E + RS003A   Tops Code: RSR   ZDV +

KDB931617+(W)   B931747 (W)

Number Series: B931750 - B931974

Description: 22T Plate Wagon
Builder: BR (Shildon Works)   Lot No.: 2476
Diagram No.: 1/431   Built: 1953
Tare Weight: 9.5t +   9.8t   10.8t *   G.L.W.: 9.8t :   32.0t +
Design Code: RS003A :   SP004G *   ZV044F   32.3t   33.5t *
             ZV173A +   Tops Code: RSR :   SPV *
                                                       ZVO +   ZVW

KDB931777 (W) ADB931807+(W)   B931823*(W)   B931911:(W)

Number Series: B931975 - B932824

Description: 22T Plate Wagon
Builder: BR (Shildon Works)   Lot No.: 2604
Diagram No.: 1/431   Built: 1954
Tare Weight: 9.5t -   9.8t %   10.0t *   G.L.W.: 9.8t %   13.0t >
             10.5t + 10.8t !   11.0t   30.0t = 33.5t
             13.0t >   32.5t +-*
Design Code: RS003A %   SP004G !   ZD087A =   Tops Code: RSR % SPV !
             ZD087E + ZS014C >   ZV044A -   ZDO =   ZDV +
             ZV044B *   ZV044D # ZV044E   ZSP >   ZVO -
                                                       ZVV

```
 B932010%(W)  KDB932193*(W)  KDB932319 (W)  KDB932431+(W)  KDB932526!(W)
TDB932030=(W)    B932195!(W)  KDB932335-     KDB932475+(W)  KDB932697+
KDB932070*(W)  KDB932228>     KDB932401+(W)  KDB932502#     KDB932824 (W)
 B932164!(W)
```

Number Series: B932825 - B933374

```
Description: 22T Plate Wagon
Builder: BR (Shildon Works)              Lot No.: 2734
Diagram No.: 1/432                       Built: 1955
Tare Weight: 9.5t    9.85t - 9.9t >      G.L.W.: 15.0t ^   32.0t
             15.0t ^                              32.5t +
Design Code: SP011A -   ZD127A    ZD127D !   Tops Code: SPV -   ZDV
             ZR182A %   ZS325A ^  ZV195A *              ZRV %   ZSV ^
             ZV195B +   ZV195C #  ZX165A >              ZXV >   ZVV *+#
```

```
KDB932889#(W)  KDB933030>(W)  KDB933157 (W)   B933248-(W)  KDB933336*(W)
KDB932893 (W)  KDB933100+(W)  DB933163%      KDB933332!(W) ADB933342^(W)
```

Number Series: B933375 - B933874

```
Description: 22T Plate Wagon
Builder: BR (Shildon Works)              Lot No.: 2862
Diagram No.: 1/432                       Built: 1956
Tare Weight: 9.5t    9.85t %  12.0t +    G.L.W.: 12.0t +   32.0t
                                                 32.5t %#!
Design Code: SP011D %   ZD127A #  ZD127D !   Tops Code: SPV %   ZDV #!
             ZR182C =   ZS114A +  ZV195A *              ZRV =   ZSV +
             ZV195C :   ZV195D <  ZX165D                ZXV     ZVV <:*
```

```
KDB933468!(W)  KDB933539!     KDB933629#(W) KDB933640*(W)  KDB933747!(W)
KDB933513<(W)   DB933567+(W)  KDB933633:(W)  B933663%(W)  CDB933768=(W)
KDB933529 (W)
```

Number Series: B933875 - B934024

```
Description: 42T Bogie Bolster Wagon
Builder: Teesside Bridge Engineering Ltd  Lot No.: 3128
Diagram No.: 1/476                        Built: 1957
Tare Weight: 9.5t    9.9t #  10.0t +      G.L.W.: 32.0t    32.5t +
Design Code: ZD127D :   ZV195B +  ZV195D *   Tops Code: ZDV :   ZVV *+
             ZX165B #                                   ZXV #
```

```
KDB933899+(W)  KDB933924*(W)  KDB933969#(W) KDB933995:(W)  KDB933998:(W)
KDB933905+(W)  KDB933947      KDB933993:(W)
```

Number Series: B934025 - B935524

```
Description: 22T Plate Wagon
Builder: BR (Shildon Works)                  Lot No.: 3223
Diagram No.: 1/434                            Built: 1959
Tare Weight: 9.5t b   10.0t #                 G.L.W.: 11.6t <^   32.0t b
             10.5t    11.6t <^                         33.0t      35.0t #
Design Code: RR001G ^   SR001A a   SR001B #   Tops Code: SRV #a    RRV ^
             ZD127B +   ZD127C %   ZD127D b              ZDV b+%   ZEB <
             ZE003A <   ZV052B     ZV052C *              ZVV       ZVW =
             ZV052E >   ZV052F -   ZV052G =              ZYW "     ZXW !
             ZY122A "   ZX165E !              Fishkind: "BREAM" <-=%"!
```

```
KDB934039>(W)  KDB934456-    KDB934788*    KDB935179*(W)  KDB935336*
KDB934064+     KDB934598 (W) KDB934794%(W)  B935182#(W)    DB935345%(W)
 B934079#(W)   KDB934599%(W)  B934881#(W)  DBB935184%(W)  KDB935371*
KDB934104+(W)   B934628#(W)  KDB934968+(W) KDB935197*(W)   DB935383%
 B934109#(W)   KDB934657 (W) KDB934983+(W)  B935245#(W)    DB935416<(W)
KDB934113 (W)  KDB934663+(W) KDB935084>(W) KDB935265-(W)   B935420#(W)
KDB934128 (W)   B934718a(W)  KDB935085>(W) KDB935275+(W)  KDB935425*
 DB934398<     KDB934720+(W) KDB935110+    KDB935277       KDB935448*(W)
 B934399^       B934731a(W)  KDB935126%(W) LDB935302"(W)  KDB935462*(W)
KDB934409+(W)  KDB934767>(W) KDB935131 (W) ADB935320=     LDB935470"(W)
KDB934427+(W)  KDB934786!(W) KDB935142!(W) KDB935322*(W)  KDB935492b(W)
 B934448#(W)
```

Number Series: B935525 - B936524

```
Description: 22T Plate Wagon
Builder: BR (Shildon Works)                  Lot No.: 3338
Diagram No.: 1/434                            Built: 1961
Tare Weight: 9.5t +   10.0t ab   10.3t c<    G.L.W.: 10.5t #   11.0t >
             10.5t    11.0t >    11.6t :             11.6t :   13.0t ^
             13.0t ^                                 32.0t +   32.5t yz
Design Code: SP011B <   SP011E c   SR001A a   SR001B b   ZD127C +    33.0t   35.0t ab
             ZD127D y   ZD127E z   ZE003A :   ZS010A >   ZS015A ^
             ZV052B =   ZV052C *   ZV052F     ZV052H -   ZY106A #
             ZD127D y   ZD127E z              Fishkind: "BREAM" :
Tops Code: SPV <c   SRV ab   ZDV +yz   ZEB :   ZSV >^   ZVV   ZVW -   ZYV #
```

```
KDB935532+(W)  KDB935920*(W)   B936062b(W)   B936192b(W)  KDB936347+(W)
KDB935556*      B935955b(W)  LDB936082#    KDB936193*(W)  KDB936348*(W)
KDB935634+(W)   B935976b(W)  KDB936098+(W) KDB936201*(W)  KDB936373 (W)
KDB935669+(W)   B935979b(W)   B936104a(W)  KDB936227=      DB936394^(W)
KDB935750+(W)  KDB935986=    KDB936123+(W)  DB936228z     TDB936402^(W)
 B935754b(W)   KDB935993=(W)  DB936129:    KDB936295*(W)   DB936405>(W)
 DB935793:      B936009b(W)  KDB936149y(W) KDB936307*(W)  KDB936443+
 B935845b(W)   ADB936011x(W)  B936159a(W)   B936322c(W)   KDB936456+(W)
KDB935862-(W)  KDB936033 (W) KDB936176*(W)  DB936323+(W)  KDB936472<(W)
KDB935888*(W)   B936048a(W)  KDB936177*(W) KDB936337+(W)   DB936524+(W)
 B935893b(W)
```

Number Series: B940050 - B940999

Description: 30T Bogie Bolster Wagon
Builder: Metropolitan Cammell Co Ltd   Lot No.: 2308
Diagram No.: 1/471   Built: 1951
Tare Weight: 14.5t -   15.5t   G.L.W.: 45.0t -   46.0t
Design Code: YN006B * YV023C + YV040C #   Tops Code: YNP * YVV +
             YY011B -   YVW # YYP -

KDB940158+(W) KDB940252+   LDB940385-(W) DB940813#(W) KDB940938+(W)
LDB940186-(W) LDB940259-(W) KDB940590*   KDB940867+(W) KDB940968*(W)

Number Series: B941555 - B941629

Description: 42T Bogie Bolster Wagon
Builder: BR (Lancing Works)   Lot No.: 2487
Diagram No.: 1/472   Built: 1953
Tare Weight: 16.0t   G.L.W.: 62.0t
Design Code: YR018A   Tops Code: YRO

CDB941616

Number Series: B941780 - B941929

Description: 42T Bogie Bolster Wagon
Builder: Cravens R C & W Co Ltd   Lot No.: 2624
Diagram No.: 1/472   Built: 1954
Tare Weight: 16.0t   G.L.W.: 62.0t
Design Code: YR018A   Tops Code: YRO

CDB941793 (W) CDB941929 (W)

Number Series: B943000 - B943099

Description: 30T Bogie Bolster Wagon
Builder: BR (Swindon Works)   Lot No.: 2326
Diagram No.: 1/471   Built: 1952
Tare Weight: 15.5t   G.L.W.: 46.0t
Design Code: YV023C   Tops Code: YVV

KDB943061

Number Series: B943100 - B943349

Description: 30T Bogie Bolster Wagon
Builder: Metropolitan Cammell Co Ltd   Lot No.: 2406
Diagram No.: 1/471   Built: 1952
Tare Weight: 15.5t +   32.0t   G.L.W.: 32.0t   46.0t +
Design Code: YV023C + YS001A   Tops Code: YSO   YVV +

KDB943124+(W) ADB943235 (W)

Number Series: B943350 - B943509

Description: 30T Bogie Bolster Wagon
Builder: BR (Swindon Works)   Lot No.: 2496
Diagram No.: 1/473   Built: 1953
Tare Weight: 23.6t   G.L.W.: 36.6t
Design Code: YY012B   Tops Code: YYV

LDB943388 (W)

Number Series: B943510 - B943609

Description: 30T Bogie Bolster Wagon   Lot No.: 2539
Builder: Metropolitan Cammell Co Ltd   Built: 1953
Diagram No.: 1/473   G.L.W.: 36.6t *:   46.5t
Tare Weight: 16.0t   23.6t *:   Tops Code: YVV   YYV *
Design Code: YV052A YY012B * YY012D :   YYW :

KDB943543   KDB943549 (W) LDB943572*(W) KDB943582   LDB943591:
KDB943545   LDB943569*(W)

Number Series: B943610 - B943859

Description: 30T Bogie Bolster Wagon
Builder: BR (Swindon Works)   Lot No.: 2542
Diagram No.: 1/473   Built: 1954
Tare Weight: 23.6t   G.L.W.: 36.6t
Design Code: YY012B   YY012D * YY012F :   Tops Code: YYV   YYW *:

LDB943627:   LDB943683 (W) LDB943700 (W) LDB943727*

Number Series: B943860 - B944299

Description: 30T Bogie Bolster Wagon   Lot No.: 2583
Builder: Metropolitan Cammell Co Ltd   Built: 1954
Diagram No.: 1/475   G.L.W.: 20.0t =   22.0t #
Tare Weight: 16.0t   17.2t # 23.6t +   36.6t +   46.5t
Design Code: YS020A = YS023B # YV052A *   Tops Code: YSR #   YSO =
            YY012A - YY012B +   YVV *   YYV

LDB943863+(W) DB943902=(W) LDB944040+(W) DB944054   DB944250#
KDB943887*   LDB943910+

Number Series: B944310 - B944469

Description: 30T Bogie Bolster Wagon
Builder: Birmingham R C & W Co Ltd   Lot No.: 2616
Diagram No.: 1/475   Built: 1954
Tare Weight: 16.0t *   23.6t   G.L.W.: 36.6t   46.5t *
Design Code: YY012B   YV052A *   Tops Code: YVV *   YYV

LDB944338 (W) KDB944340*(W) LDB944359 (W) LDB944427 (W) LDB944438 (W)

Number Series: B945390 - B945790

Description: 30T Bogie Bolster Wagon
Builder: Metropolitan Cammell Co Ltd   Lot No.: 2818
Diagram No.: 1/475   Built: 1956
Tare Weight: 17.5t   G.L.W.: 48.0t
Design Code: YY033A   Tops Code: YYV

LDB945668 (W)

Number Series: B945791 - B945990

Description: 30T Bogie Bolster Wagon   Lot No.: 3155
Builder: Metropolitan Cammell Co Ltd   Built: 1958
Diagram No.: 1/477   G.L.W.: 36.6t :!   48.5t
Tare Weight: 18.0t   23.6t :!   Tops Code: BCV =   YNV -
Design Code: BC003C = YN048A - YV057C *   YVV *   YVW #
             YV057J # YX035A + YY012E :   YXV +   YYV
             YY012G ! YY036B   YYW :!

KDB945798+   LDB945850   B945878=(W)   B945925=(W) LDB945931:(W)
KDB945816*(W) KDB945859-   KDB945885-(W) LDB945927:(W) LDB945961 (W)
  B945827=(W)   B945860=(W) ADB945913#(W) KDB945928*   B945979=(W)
KDB945841-(W) LDB945865:

Number Series: B947860 - B948009

Description: 42T Bogie Plate Wagon
Builder: BR (Ashford Works)   Lot No.: 3229
Diagram No.: 1/492   Built: 1958
Tare Weight: 15.5t   17.0t *#% 19.5t -!   G.L.W.: 58.0t   62.0t -!
Design Code: YS027C   YS027D * YS027E :   63.0t *#%^
             YS027F = YS027G # YS039B !   Tops Code: YSV =^   YSW
             YS040A ^ YS040C -

KDB947863   KDB947891!   KDB947920!(W) KDB947951   KDB947981*
KDB947864 (W) KDB947894   KDB947922   KDB947952-(W) KDB947982 (W)
KDB947865   KDB947895:   KDB947923!   KDB947953*(W) KDB947985 (W)
KDB947866*   KDB947896!   KDB947924   KDB947954   KDB947986
KDB947869   KDB947897 (W) KDB947925   KDB947955   KDB947987*
KDB947870 (W) KDB947898   KDB947926   KDB947956   KDB947988*
KDB947872   KDB947900   KDB947928!(W) KDB947958!   KDB947989*
KDB947873   KDB947901:   KDB947929 (W) KDB947962:   KDB947990
KDB947874   KDB947902:   KDB947930!   KDB947964!   KDB947995 (W)
KDB947875   KDB947903 (W) KDB947931   KDB947965*   KDB947997 (W)
KDB947877   KDB947905!   KDB947932 (W) KDB947966!   KDB947998-(W)
KDB947878   KDB947907*   KDB947936=   KDB947967   KDB948000^(W)
KDB947881   KDB947908*   KDB947938 (W) KDB947969^   KDB948001^(W)
KDB947882   KDB947909*(W) KDB947939 (W) KDB947970-(W) KDB948003-(W)
KDB947883!   KDB947910!   KDB947940   KDB947971*   KDB948004 (W)
KDB947884   KDB947911   KDB947941   KDB947973-   KDB948005-(W)
KDB947885 (W) KDB947914   KDB947942%   KDB947976*   KDB948006 (W)
KDB947887!   KDB947915:   KDB947943!   KDB947977-(W) KDB948008 (W)
KDB947888*   KDB947919!   KDB947944   KDB947978 (W) KDB948009 (W)
KDB947889!

Number Series: B948010 - B948134

Description: 42T Bogie Plate Wagon   Lot No.: 3235
Builder: BR (Derby Works)   Built: 1959
Diagram No.: 1/490   G.L.W.: 62.0t
Tare Weight: 19.5t   19.6t *   Tops Code: BPV *   YRV +
Design Code: BP001B *   YR026A +   YU001A #   YUV #   YVV =%
             YV024B =   YV024G %   YY041A   YYV

```
LDB948021      KDB948048%(W)     B948072*(W)  KDB948106=    KDB948126#(W)
KDB948032=(W)   B948052*(W)  KDB948073+        B948118*(W) KDB948133=
KDB948044+     KDB948059#(W)
```

Number Series: B948135 - B948259

Description: 42T Bogie Plate Wagon
Builder: BR (Derby Works)   Lot No.: 3236
Diagram No.: 1/490   Built: 1960
Tare Weight: 19.5t   19.6t %   G.L.W.: 62.0t
Design Code: BP001B %   YU001A +   YU001B *   Tops Code: BPV %   YUV +
             YX057A   YUW *   YXV !

```
KDB948150+(W) KDB948168+(W)  KDB948186+(W) KDB948196+(W)    B948229%(W)
 B948151*(W)   B948174%(W)  KDB948187=(W)   B948222%(W) KDB948259*
KDB948152     B948178%(W)  KDB948195*(W)   B948228%(W)
```

Number Series: B948260 - B948409

Description: 42T Bogie Plate Wagon
Builder: BR (Ashford Works)   Lot No.: 3240
Diagram No.: 1/490   Built: 1959
Tare Weight: 16.0t *   19.5t   26.0t +   G.L.W.: 45.0t +   58.5t *
Design Code: BP002B %   YN058A =   YS039C *   62.0t
             YU002A !   YV067A #   YV067B ^
             YX056A   YY040B +
Tops Code: BPV %   YNV =   YSW *   YUV !   YVV #   YXV   YYW +

```
LDB948260+     KDB948293      LDB948316+    LDB948347+    LDB948385+
KDB948268*     LDB948294+      B948319%(W) KDB948354!    LDB948392+
LDB948270+      B948300 (W) LDB948321+     LDB948360+    LDB948398+
LDB948278+     KDB948307!(W) LDB948326+    KDB948362*    ADB948399#(W)
KDB948280!(W) LDB948311+     LDB948328+     B948366%(W) RDB948407=(W)
LDB948282+     LDB948314+     LDB948334+    LDB948375+    LDB948409+
LDB948286+     LDB948315+    KDB948337^(W) LDB948377+
```

Number Series: B949050 - B949089

Description: 56T Strip Coil Wagon
Builder: Head Wrightson & Co Ltd   Lot No.: 3015
Diagram No.: 1/404   Built: 1957
Tare Weight: 23.25t   G.L.W.: 84.0t
Design Code: BV001B   Tops Code: BVW

```
  B949051 (W)     B949054 (W)    B949055 (W)    B949057       B949058 (W)
```

```
B949059 (W)      B949066 (W)      B949073 (W)      B949078 (W)      B949085 (W)
B949060 (W)      B949067 (W)      B949074 (W)      B949079 (W)      B949086 (W)
B949061 (W)      B949068 (W)      B949075 (W)      B949080 (W)      B949087 (W)
B949062          B949069 (W)      B949076 (W)      B949081 (W)      B949088 (W)
B949063 (W)      B949071 (W)      B949077 (W)      B949083 (W)      B949089 (W)
B949064 (W)      B949072 (W)
```

Number Series: B949133 – B949179

Description: 21T Strip Coil Wagon
Builder: BR (Derby Works)   Lot No.: 3450
Diagram No.: 1/412   Built: 1962
Tare Weight: 11.0t   G.L.W.: 32.5t
Design Code: SF002B   Tops Code: SFW

```
B949133 (W)      B949142 (W)      B949152 (W)      B949164 (W)      B949173 (W)
B949134 (W)      B949143 (W)      B949154 (W)      B949165 (W)      B949174 (W)
B949135 (W)      B949144 (W)      B949155 (W)      B949166 (W)      B949175 (W)
B949138 (W)      B949145 (W)      B949156 (W)      B949168 (W)      B949176 (W)
B949140 (W)      B949146 (W)      B949160 (W)      B949171 (W)      B949178 (W)
B949141 (W)      B949150 (W)      B949163 (W)
```

Number Series: B949180 – B949185

Description: 24T Strip Coil Wagon
Builder: BR (Derby Works)   Lot No.: 3464
Diagram No.: 1/414   Built: 1962
Tare Weight: 11.5t   G.L.W.: 36.0t
Design Code: SG001B   Tops Code: SGW

```
B949180 (W)      B949181 (W)      B949183 (W)      B949184 (W)      B949185 (W)
```

Number Series: B949186 – B949197

Description: 24T Strip Coil Wagon
Builder: BR (Derby Works)   Lot No.: 3478
Diagram No.: 1/414   Built: 1963
Tare Weight: 11.5t   G.L.W.: 36.0t
Design Code: SG001B   Tops Code: SGW

```
B949188 (W)      B949192 (W)      B949193 (W)      B949194 (W)      B949195 (W)
```

Number Series: B949198 – B949209

Description: 24T Strip Coil Wagon
Builder: BR (Derby Works)   Lot No.: 3513
Diagrom No.: 1/414   Built: 1964
Tare Weight: 11.5t   G.L.W.: 36.0t
Design Code: SG001B   Tops Code: SGW

```
B949198 (W)      B949200 (W)      B949203 (W)      B949205 (W)      B949208 (W)
B949199 (W)      B949202 (W)      B949204 (W)      B949206 (W)      B949209 (W)
```

Number Series: B949210 - B949219

Description: 24T Strip Coil Wagon
Builder: BR (Derby Works)
Diagram No.: 1/414
Tare Weight: 11.5t
Design Code: SG001B

Lot No.: 3514
Built: 1964
G.L.W.: 36.0t
Tops Code: SGW

    B949212 (W)   B949214 (W)   B949215 (W)   B949217 (W)   B949219 (W)

Number Series: B949503 - B949550

Description: 57T Slab Wagon
Builder: BR (Swindon Works)
Diagram No.: 1/408   1/417   1/419
Tare Weight: 21.45t
Design Code: BN001A

Lot No.: 3359
Built: 1961
G.L.W.: 82.5t
Tops Code: BNX

    B949503   B949517   B949536   B949541   B949548
    B949508   B949519 (W)   B949537   B949542   B949549
    B949509   B949525   B949539   B949545   B949550
    B949512   B949535

Number Series: B950000 - B950124

Description: 20T Goods Brake Van
Builder: BR (Derby Works)
Diagram No.: 1/503
Tare Weight: 20.5t
Design Code: CA004B -   ZT005A +   ZT005B
             ZT005C *

Lot No.: 2025
Built: 1950-51
G.L.W.: 20.5t
Tops Code: CAP -   ZTO
           ZTP *+

    DB950001+(W)   DB950019 (W)   DB950047 (W)   B950055-(W)   KDB950098*(W)
    KDB950006*(W) DB950020 (W)   DB950048 (W)   DB950076*(W) DB950104 (W)
    DB950007 (W)   B950039-(W)

Number Series: B950125 - B950249

Description: 20T Goods Brake Van
Builder: BR (Derby Works)
Diagram No.: 1/505
Tare Weight: 20.5t
Design Code: ZT003A   ZT003B -   ZT003C +

Lot No.: 2026
Built: 1950
G.L.W.: 20.5t
Tops Code: ZTO   ZTP -
           ZTR +

    DB950133 (W)   KDB950159-   KDB950205-(W) KDB950213-   LDB950227-(W)
    DB950134 (W)   DB950161 (W)   DB950206 (W)   DB950215 (W) DB950231 (W)
    DB950135 (W)   DB950180 (W)   DB950209 (W)   DB950225 (W) DB950236 (W)
    KDB950155+(W) DB950196 (W)

Number Series: B950250 - B950539

Description: 20T Goods Brake Van
Builder: BR (Darlington Works)
Diagram No.: 1/500
Tare Weight: 20.5t
Design Code: ZT001A + ZT001B

Lot No.: 2051
Built: 1949
G.L.W.: 20.5t
Tops Code: ZTO + ZTV

DB950269+(W)   DB950307+(W)   DB950344+(W)   DB950406+(W)   DB950416+(W)
DB950303 (W)   DB950319+(W)

Number Series: B950542 - B950615

Description: 20T Goods Brake Van
Builder: BR (Swindon Works)
Diagram No.: 1/502
Tare Weight: 20.5t
Design Code: ZP001B   ZP001D +

Lot No.: 2099
Built: 1949-50
G.L.W.: 20.5t
Tops Code: ZPP   ZPR +

DB950551+   DB950606 (W)

Number Series: B950616 - B950865

Description: 20T Goods Brake Van
Builder: BR (Darlington Works)
Diagram No.: 1/504
Tare Weight: 20.5t
Design Code: CA003D :   ZT004A +   ZT004C -
             ZT004H *   ZT004F

Lot No.: 2136
Built: 1950
G.L.W.: 20.5t
Tops Code: CAP :   ZTO +
           ZTP -   ZTR

DB950619+(W)   B950694:(W) KDB950738-   KDB950787*(W) DB950817+(W)
DB950638+(W)   DB950710+(W)   DB950746+(W) LDB950790   DB950820+(W)
TDB950657-(W) KDB950718-(W)   DB950750+(W) KDB950815-(W) DB950850+(W)
DB950661+   LDB950728 (W)   DB950775+(W)

Number Series: B950866 - B951115

Description: 20T Goods Brake Van
Builder: BR (Darlington Works)
Diagram No.: 1/506
Tare Weight: 20.0t   20.5t +
Design Code: ZT006A + ZT006K

Lot No.: 2137
Built: 1950-51
G.L.W.: 20.0t   20.5t +
Tops Code: ZTO + ZTP

KDB950878 (W)   DB950882 (W)   DB950954+(W) DB951040+(W)   DB951042+(W)
DB950880+(W)   DB950912+(W)   DB950987+(W)

Number Series: B951116 - B951275

Description: 20T Goods Brake Van
Builder: BR (Darlington Works)
Diagram No.: 1/504
Tare Weight: 20.5t
Design Code: ZT004A + ZT004C * ZT004F

Lot No.: 2206
Built: 1951
G.L.W.: 20.5t
Tops Code: ZTO + ZTP *
          ZTR

DB951138+(W)	KDB951156:(W)	DB951205+(W)	DB951246+(W)	DB951263+(W)
DB951148+(W)	LDB951175	DB951218+(W)	DB951260*(W)	DB951275+(W)
DB951150+(W)	DB951176+(W)			

Number Series: B951276 - B951515

Description: 20T Goods Brake Van
Builder: BR (Darlington Works)
Diagram No.: 1/506
Tare Weight: 20.5t
Design Code: ZT004A + ZT004C

Lot No.: 2207
Built: 1951
G.L.W.: 20.5t
Tops Code: ZTO + ZTP

DB951283 (W)	KDB951323 (W)	DB951350 (W)	DB951428+(W)	KDB951455
DB951294+(W)	DB951338+(W)	DB951404+(W)	DB951433+(W)	DB951474+(W)
DB951322+(W)	DB951349+(W)	KDB951425+(W)	DB951449+(W)	DB951491+(W)

Number Series: B951516 - B951715

Description: 20T Goods Brake Van
Builder: BR (Darlington Works)
Diagram No.: 1/506
Tare Weight: 20.0t * 20.5t
Design Code: ZT006A   ZT006K *

Lot No.: 2349
Built: 1952-53
G.L.W.: 20.0t * 20.5t
Tops Code: ZTO   ZTP *

DB951522 (W)	DB951624 (W)	DB951693 (W)	DB951708*	DB951710 (W)
DB951618 (W)	DB951675 (W)			

Number Series: B951716 - B951865

Description: 20T Goods Brake Van
Builder: BR (Darlington Works)
Diagram No.: 1/506
Tare Weight: 20.0t %   20.5t
Design Code: CA005E *   ZT006A +   ZT006M %
             ZT006V

Lot No.: 2350
Built: 1952
G.L.W.: 20.0t %   20.5t
Tops Code: CAP *   ZTO +
           ZTP   ZTR %

KDB951754 (W)	LDB951762	DB951777+(W)	DB951813+(W)	KDB951843+(W)
DB951759+(W)	DB951773+(W)	B951782*(W)	DB951815+(W)	

Number Series: B951866 - B952005

Description: 20T Goods Brake Van
Builder: BR (Darlington Works)   Lot No.: 2477
Diagram No.: 1/506   Built: 1953
Tare Weight: 20.0t %   20.5t   G.L.W.: 20.0t %   20.5t
Design Code: ZT006A +   ZT006E   ZT006K %   Tops Code: ZTO +   ZTP

    DB951866+(W)   DB951879+(W)   DB951913+(W)   DB951940+(W)   DB951968!(W)
    DB951878 (W) KDB951885%(W)   DB951915 (W)

Number Series: B952006 - B952165

Description: 20T Goods Brake Van
Builder: BR (Darlington Works)   Lot No.: 2478
Diagram No.: 1/506   Built: 1953
Tare Weight: 20.5t   G.L.W.: 20.5t
Design Code: CA005A *   ZT006A   Tops Code: CAO *   ZTO

    DB952025 (W)   B952078*(W)   DB952106 (W)   DB952127 (W)   DB952133 (W)

Number Series: B952166 - B952515

Description: 20T Goods Brake Van
Builder: BR (Darlington Works)   Lot No.: 2605
Diagram No.: 1/506   Built: 1954
Tare Weight: 20.0t   20.5t +!*   G.L.W.: 20.0t   20.5t +!*
Design Code: CA005A -   CA005E =   ZT006A + Tops Code: CAP =   CAO -
             ZT006E !   ZT006K %   ZT006M #   ZSQ   ZTO +
             ZT006W *   ZS128C   ZTR #*   ZTP !%

    DB952212%   DB952279+(W)   DB952300+(W)   DB952323+   LDB952472#
    DB952229+   DB952281+(W)   B952302=(W)   DB952399!(W)   DB952482+(W)
    DB952267+(W) ADB952282%   LDB952317*(W)   DB952429+(W)   DB952505%
    DB952273+(W) ADB952294 (W)   DB952322+(W)   B952445=(W)   B952511-(W)

Number Series: B952516 - B952715

Description: 20T Goods Brake Van
Builder: BR (Darlington Works)   Lot No.: 2606
Diagram No.: 1/506   Built: 1954-55
Tare Weight: 20.5t   G.L.W.: 20.5t
Design Code: ZT006A   ZT006C *   Tops Code: ZTO   ZTV *

    DB952540 (W)   DB952610   ADB952647 (W)   DB952674 (W)   DB952693 (W)
    DB952580 (W)   DB952621 (W)   DB952654*   DB952675*   DB952710 (W)
    DB952602 (W)   DB952638 (W)   DB952665 (W)

Number Series: B952716 - B953115

```
Description: 20T Goods Brake Van          Lot No.: 2741
Builder: BR (Darlington Works)           Built: 1955
Diagram No.: 1/504                       G.L.W.: 20.5t
Tare Weight: 20.5t                       Tops Code: ZTO    ZTP *
Design Code: ZT004A    ZT004C *  ZT004G =              ZTQ =
```

```
DB952726 (W)    DB952818 (W)    DB952932 (W)    DB952985 (W)    DB953045 (W)
DB952737 (W)    DB952827 (W)    ADB952946=(W)   DB952998 (W)    DB953053 (W)
DB952759 (W)    DB952834 (W)    DB952970 (W)    KDB952999*      DB953101 (W)
DB952786 (W)    DB952868 (W)    DB972973 (W)    DB953009 (W)    DB953114 (W)
DB952810 (W)    DB952907 (W)    DB952975 (W)    DB953033 (W)
```

Number Series: B953116 - B953415

```
Description: 20T Goods Brake Van
Builder: BR (Darlington Works)           Lot No.: 2868
Diagram No.: 1/506                       Built: 1956-57
Tare Weight: 20.0t !    20.5t            G.L.W.: 20.0t !    20.5t
Design Code: ZS128D !  ZT006B   ZT006D * Tops Code: ZSQ !   ZTO
             ZT006Q =                               ZTR =   ZTV *
```

```
DB953153        DB953180 (W)    DB953231        DB953310 (W)    ADB953398!(W)
DB953154 (W)    DB953194 (W)    DB953250 (W)    DB953353 (W)    DB953399*
DB953159 (W)    LDB953206=      DB953291 (W)    DB953381 (W)    DB953411 (W)
```

Number Series: B953416 - B953675

```
Description: 20T Goods Brake Van
Builder: BR (Darlington Works)           Lot No.: 3012
Diagram No.: 1/506                       Built: 1957-58
Tare Weight: 20.0t !    20.5t            G.L.W.: 20.0t !    20.5t
Design Code: CA005B =  ZS128D !  ZT006B  Tops Code: CAV =   ZSQ !
             ZT006R *  ZT006Q +                     ZTO     ZTR +*
```

```
DB953422 (W)    DB953485 (W)    DB953517 (W)    DB953563 (W)    DB953634 (W)
DB953431 (W)    DB953488 (W)    DB953518 (W)    DB953570 (W)    DB953637 (W)
DB953435 (W)    DB953495 (W)    DB953520 (W)    DB953580 (W)    DB953639 (W)
DB953446 (W)    LDB953497+      DB953521 (W)    DB953585 (W)    DB953640 (W)
DB953468 (W)    DB953503 (W)    LDB953547*      DB953599 (W)    LDB953643 (W)
DB953475 (W)    DB953508 (W)    DB953552 (W)    LDB953602+      DB953656=(W)
DB953480 (W)    ADB953515!      DB953553 (W)    DB953630 (W)    DB953660 (W)
```

Number Series: B953676 - B954520

```
Description: 20T Goods Brake Van
Builder: BR (Darlington Works)           Lot No.: 3129
Diagram No.: 1/506                       Built: 1958-59
Tare Weight: 20.0t <z   20.5t            G.L.W.: 20.0t <z   20.5t
Design Code: CA005L x  ZS128D <  ZS128E > Tops Code: CAR x  ZSQ ><
             ZT006B +  ZT006D    ZT006F *            ZTP *  ZTO +-%
             ZT006G -  ZT006L %  ZT006Q ^            ZXO z  ZTR !=^
             ZT006R !  ZT006S #  ZT006T y            ZTV    ZTQ #y
             ZT006U =  ZX120B z  ZX120C :            ZXQ :
```

DB953676+(W)	DB953827	DB954050+(W)	DB954210+(W)	DB954373+(W)
DB953677+(W)	DB953831+(W)	DB954059+(W)	DB954215+(W)	DB954377+(W)
DB953678+(W)	DB953835+(W)	ADB954061z(W)	LDB954219^	DB954381+(W)
DB953683+(W)	DB953840+(W)	DB954062+(W)	DB954221+(W)	DB954389+(W)
DB953691+(W)	DB953848+(W)	DB954066+(W)	DB954224+(W)	DB954391+(W)
DB953694:	DB953854+(W)	DB954084+(W)	DB954230+(W)	DB954392+(W)
DB953696+(W)	DB953861+(W)	DB954088+(W)	DB954231+(W)	DB954403+(W)
DB953697!	DB953867+(W)	DB954090+(W)	DB954236+(W)	LDB954406^
DB953701*(W)	DB953868+(W)	DB954103+(W)	DB954241+(W)	KDB954429=(W)
DB953711+(W)	DB953870"(W)	DB954104+(W)	LDB954256y(W)	DB954430+(W)
DB953716+(W)	LDB953889^	DB954112+(W)	DB954285+(W)	B954432+(W)
DB953717+(W)	DB953890+(W)	DB954122+(W)	LDB954300^	B954433z
DB953737+(W)	DB953896+(W)	DB954126+(W)	DB954301+(W)	DB954453+(W)
DB953741+(W)	DB953898+(W)	DB954127+(W)	DB954302+(W)	DB954455+(W)
LDB953745%(W)	LDB953899%(W)	DB954130+(W)	DB954306+(W)	DB954471+(W)
DB953755+(W)	DB953938+(W)	ADB954145z(W)	DB954313+(W)	DB954473+(W)
DB953757+(W)	ADB953939>	DB954159+(W)	DB954318+(W)	LDB954476+(W)
DB953758+(W)	DB953954+(W)	KDB954164=	DB954319+(W)	DB954478+(W)
DB953762+(W)	ADB954014<	KDB954173=	DB954340+(W)	DB954479+(W)
LDB953787#(W)	DB954015+(W)	DB954182+(W)	DB954347+(W)	DB954501+(W)
DB953788+(W)	DB954021+(W)	DB954184+(W)	DB954348+(W)	ADB954510+(W)
KDB953793=(W)	DB954030+(W)	DB954190+(W)	DB954353+(W)	DB954511+(W)
DB953804+(W)	DB954039+(W)	DB954199	DB954358+(W)	DB954514+(W)
DB953813+(W)	DB954047+(W)	DB954208+(W)	DB954368+(W)	DB954518+(W)

Number Series: B954521 - B954997

Description: 20T Goods Brake Van	Lot No.: 3227
Builder: BR (Darlington Works)	Built: 1959
Diagram No.: 1/507	G.L.W.: 19.0t -
Tare Weight: 19.0t - 20.0t #=z: 20.5t	20.0t #=z: 20.5t

Design Code: CA006C =   CA006D -   RA001A !
            RF002B -   ZP047A   ZS151A %
            ZT007A *   ZT007B <   ZT007C y
            ZT007D #   ZT007E ^   ZT007F z
            ZT007F z   ZT007H +   ZT007J >

Tops Code: CAP =   CAR :
          RAR !   RFQ -
          ZPR   ZSQ %
          ZTO *z   ZTP <^
          ZTR y#+   ZTV >

B954522:(W)	B954581:	B954645:	ADB954706<(W)	DB954765*(W)
B954524-	LDB954589^	DB954650<	B954709:	B954767-
KDB954528y	LDB954595#	DB954651+(W)	LDB954716+	B954768:
B954531:(W)	DB954595+	DB954652+	DB954718+(W)	B954769-
B954532:	DB954597+	DB954659*(W)	B954723:	DB954772*(W)
DB954534*(W)	DB954598*(W)	DB954660*(W)	B954726:	B954779:
B954536:	B954599:(W)	B954661:(W)	B954734:	B954781:
DB954537+(W)	B954600:	DB954671*(W)	B954735:	B954782:
B954544:(W)	DB954605+(W)	DB954672+	B954739:	B954786:
B954549:(W)	LDB954609+	B954673:	B954740:	B954788:
ADB954550+	DB954612:(W)	DB954674<(W)	KDB954747+(W)	B954791=(W)
ADB954551%	B954613:	B954678:(W)	DB954748*(W)	B954793:
B954552!	B954616:	B954682-	B954750:	LDB954794+
B954561=	B954617:(W)	ADB954685y	B954751:	DB954797+
DB954564^(W)	B954620-	B954686:	B954753:	DB954798<(W)
DB954565*	B954622:(W)	DB954688+	DB954756*(W)	B954802:(W)
B954566-	DB954624*(W)	B954690:(W)	B954759:	DB954806*(W)
B954567:(W)	B954625:	B954692:	ADB954760y	B954807:
B954568:(W)	B954627:	B954693:	LDB954762+	DB954809+(W)
DB954569*(W)	KDB954630+	B954702:(W)	B954763:	B954812:
DB954575*(W)	LDB954636+	DB954703*(W)	DB954764*(W)	B954813-

LDB954815y(W)	B954860-	B954893:(W)	B954928:(W)	B954977:(W)
ADB954816#	B954861:	B954894:	DB954930^(W)	B954978:(W)
DB954817+	B954862:	DB954898+	B954931:(W)	B954980:
ADB954819y	B954865:(W)	KDB954899y	B954935-	B954981:(W)
B954821:	B954871:	DB954900*(W)	ADB954937<	B954983:
B954822:	B954874:(W)	B954901:(W)	DB954943<(W)	B954985:
B954827:	B954876-	B954902:	B954944:	DB954988
B954829:(W)	DB954877+	B954908:(W)	LDB954950y(W)	B954989:(W)
DB954835*(W)	DB954878*(W)	B954911:(W)	B954952-	B954990:
B954842:	B954881:	B954912-	DB954956 (W)	B954991:(W)
DB954843+	B954885:(W)	DB954920*(W)	B954962:	B954992:
B954852:(W)	B954886:(W)	B954921:	B954964:(W)	LDB954994#
DB954855*(W)	B954890:(W)	B954925:	B954966:(W)	B954997:

Number Series: B954998 - B955247

Description: 20T Goods Brake Van
Builder: BR (Ashford Works)
Diagram No.: 1/507
Tare Weight: 20.0t -# 20.5t
Design Code: CA006C = CA006D   CA006F -
             CA006H ^ RA001A * ZT007A +
             ZT007B ! ZT007C # ZT007E <
             ZT007H >

Lot No.: 3394
Built: 1963
G.L.W.: 20.0t -#   20.5t
Tops Code: CAP = CAR
           RAR * ZTO +
           ZTP !< ZTR #>

B954998 (W)	B955043	B955086 (W)	B955139	B955187 (W)
B954999 (W)	B955045	B955087	DB955142+(W)	DB955191+(W)
DB955000+(W)	ADB955046!(W)	B955089	B955143	DB955192+(W)
DB955003+(W)	DB955047#(W)	B955090	KDB955146>	B955196
B955006	B955048 (W)	B955091 (W)	DB955147+(W)	B955197
B955007	B955049-	KDB955094#(W)	B955151	DB955200>(W)
B955008	DB955051+(W)	B955097	B955152 (W)	B955204^(W)
ADB955009#	B955052	B955101	B955153	DB955206+
B955010	DB955055!	B955104	DB955158+(W)	B955208=(W)
B955014*	DB955057!(W)	B955107	B955160	B955211 (W)
B955017 (W)	B955060	B955109	B955163	B955213 (W)
B955019 (W)	B955061 (W)	ADB955114#	KDB955166>	B955217
B955021	B955062	B955115	DB955167+	B955219 (W)
B955022 (W)	B955063	B955120-(W)	B955168	B955227
B955025	B955066	DB955121+(W)	ADB955169!(W)	B955229 (W)
B955026	DB955067#(W)	DB955124+(W)	DB955170>	B955235
B955028	DB955071#	B955125	ADB955172>	B955236
B955032	B955073	LDB955127>	DB955173>(W)	DB955237+(W)
DB955033+(W)	B955077	ADB955129>	ADB955175+(W)	DB955241>
DB955035+(W)	DB955079+(W)	DB955132>	KDB955178!(W)	DB955242+(W)
B955036	B955081 (W)	LDB955133>	DB955182+(W)	B955245!
B955037	B955082	B955134	DB955184+(W)	B955246
B955041	DB955084>(W)	B955136 (W)		

Tinsley TMD was the location for this photograph of TDB904529 a 25T
Lowmac Wagon on 11th August 1993 which has since been withdrawn.
Paul W. Bartlett

LDB924062 is a 30T Bogie Bolster Wagon modified for departmental
service. It was photographed on 3rd August 1989 at Millerhill.
Paul W. Bartlett

Withdrawn 22T Plate Wagon, KDB932824 was photographed at Bescot on 4th
September 1993.   Peter Ifold

When BR first placed orders for Brake Vans several different types
reflecting the pre-nationalisation companies designs were built.
This 20T example was based on the LMS design.   KDB950006 dating from
1950 was withdrawn at the end of 1993 and was photographed at Ditton
on 28th November 1992.   Peter Ifold

Also Available from South Coast Transport Publishing

Departmental Coaching Stock
@£6.95

British Rail Internal Users
@£7.95

R.I.V. Wagon Fleet
@£5.95

These publications and further copies of British Rail Wagon Fleet - British Railways ("B" - Prefixed Series) Freight Stock @ £6.95 per copy (all prices inc. post & packing) may be obtained from our Mail Order Dept. at the address below:-

33, Porchester Road,
Woolston,
Southampton,
Hampshire SO2 7JB

Please make cheques and postal orders payable to S.C.T. Publishing.

Trade enquiries are welcomed and these should be sent to:-

3, Morley Drive,
Bishop's Waltham,
Hampshire SO3 1RX

* * * * * * * * * *

COVER PHOTOGRAPHS

Front:-

BSW B927916, a 42T Bogie Bolster Wagon, was photographed at Mossend on 22nd July 1990 whilst engaged on timber traffic duties. It was subsequently transfered to departmental use and was withdrawn in December 1993.

Paul W. Bartlett

Rear (Upper):-

KDB740918, a 13T Pipe Wagon when in revenue service, is seen at Luton on 25th November 1990.

Paul W. Bartlett

Rear (Lower):-

Pictured at Workington on 17th August 1993 is 20T Brake Van B954781, tops code CAR.

Paul W. Bartlett